Water Resources Development and Management

Indexed by Scopus

Each book of this multidisciplinary series covers a critical or emerging water issue. Authors and contributors are leading experts of international repute. The readers of the series will be professionals from different disciplines and development sectors from different parts of the world. They will include civil engineers, economists, geographers, geoscientists, sociologists, lawyers, environmental scientists and biologists. The books will be of direct interest to universities, research institutions, private and public sector institutions, international organisations and NGOs. In addition, all the books will be standard reference books for the water and the associated resource sectors.

More information about this series at http://www.springer.com/series/7009

Naim Haie

Transparent Water Management Theory

Sefficiency in Sequity

 Springer

Naim Haie
University of Minho
Guimarães, Portugal

ISSN 1614-810X   ISSN 2198-316X (electronic)
Water Resources Development and Management
ISBN 978-981-15-6286-0   ISBN 978-981-15-6284-6 (eBook)
https://doi.org/10.1007/978-981-15-6284-6

This Springer imprint is published by the registered company Springer Nature Singapore Pte Ltd.
The registered company address is: 152 Beach Road, #21-01/04 Gateway East, Singapore 189721,
Singapore

To "Unity in Diversity"

Preface

During the past three decades, I participated and read about two intensifying story lines in water management, each with its own legitimate, albeit partial, reasons. (A) the catchphrase was 'water is life' but the focus of its management was through other valid domains, such as economics, ecosystem, food and health; (B) major integrated (water) management models and tools were developed but water inconsistencies and problems grew significantly. My sense of ambivalence grew when about a decade ago I witnessed in the Winrock International Water Forum that I had the privilege to be a member, a prolonged, intense and eventually personal exchange of ideas between two famous water scholars Dr. Peter Gleick and Dr. Chris Perry. On the subject matter at hand, they presented two separate water views, each rejecting the other passionately, and only ended when one of them left the Forum. As water problems, dichotomies and policy uncertainties grew, years ago I started writing many notes for my learning, which proved very difficult but finally lead in writing this book. My only reason to do so is the hope that it may promote, in a small way, a more sustainable and equitable water management.

All the principal ideas in this book are known; however, they are organised into a theory and an approach so that their significance can be more appreciated, while making it easier, more coherent and transparent to apply them. In travelling along this path, the book differentiates and comprehensively integrates in an explicit and objective manner many of the water concepts, such as:

- Water resources and (re)use systems
- Water management theory based on five foundational ideas
- Three water management pillars and their trade-offs
- Stakeholders, learning and smart technology
- Water management terminology and system transparency
- Sustainable equity, efficiency and conservation
- Two states of water outflow vs. one state for other resources
- Inflow and outflow efficiencies

- Water management categorisation and segmentation
- Water losses and unrecoverables at the flow and system levels
- Macro, meso and micro levels of water management
- Four types of policy in water management

In summary, this book highlights a theory that serves as a solid foundation for a comprehensive, systemic and water-centric management approach. It integrates two performance principles essential for sustainable water use systems, namely, equity and efficiency. It decreases the policy space for decision making encountered by water managers and facilitates advancing towards a reasonable and fair solution because of the bounded rationality inherent in its development. In combining the distributive and aggregative principles, the approach is at once transparent due to an autonomous structure, stakeholder enabler through learning, and technology promoter for gathering water data. These features are possible owing to the robust and comprehensive terminology that advances a unifying language for all types of water use systems, such as urban, agriculture and industry. Consequently, if a reader finds a statement less true, I suggest focusing on the words because each term has a concise meaning, requiring higher attention in reading the book.

Guimarães, Portugal   Naim Haie
April 2020

Acknowledgements

I am grateful for my learning to many distinguished water experts and decision-makers, on the one hand, and scholars and professionals in such fields as governance, social theories and ethics, on the other.

Within the last decade or so, there were students, friends and colleagues that were helpful and sometimes essential in developing this book, and as such, I am very much indebted to them. They are alphabetically: Andrew A. Keller, Asit K. Biswas, Cecilia Tortajada, Chris J. Perry, Daniel P. Loucks, David Molden, Dogan Altinbilek, Francisco N. Correia, François Molle, Gaspar J. Machado, Gordon Young, Jun Xia, Laura E. Osinska, Miguel R. Freitas, Muhammad T. Ahmad, Nader Saiedi, Patrick Lavarde, Peter H. Gleick, Raya M. Stephan, Rui M. S. Pereira, Tom Soo, Tushaar Shah.

I also would like to thank the staff at Springer, particularly Loyola D'Silva and Prasanna Kumar Narayanasamy, for their help and support.

Contents

Abbreviations and Symbols

Symbol	Description	Unit
3ME	Macro, Meso, Micro SE	%
b	Benefit index	–
bSefficiency	SE_b = beneficial Sefficiency ($W_{qx} = 1$); quantity Sefficiency	%
C	Consumption (= ET + NR)	L3; L3/L2
c	Consumptive index; OUT index	–
C1	Consumption fraction (= C/I)	–
CE	Classical Efficiency	%
cMacroSE	Consumptive $MacroSE_S$ ($ic = 0$)	%
cMacroSE$_b$	Consumptive $MacroSE_b$ ($ic = 0$)	%
cMesoSE	Consumptive $MesoSE_S$ ($ic = 0$)	%
cMesoSE$_b$	Consumptive $MesoSE_b$ ($ic = 0$)	%
cMicroSE	$= cMicroSE_S = MicroSE$	%
cSE	$= cSE_S = cSefficiency = Consumptive Sefficiency$ ($ic = 0$)	%
cSE$_b$	Consumptive bSefficiency ($ic = 0$)	%
d	Desirables; compact form of b, q or s	–
DS$_{req}$	Downstream required water	L3; L3/L2
EE	Effective Efficiency	%
ET	Evapotranspiration	L3; L3/L2
i	IN Sefficiency; IN index	–
I	Inflow (= V1 + OS + PP)	L3; L3/L2
ic	IN or inflow Sefficiency (= 1); OUT or consumptive Sefficiency (= 0)	–
iMacroSE	Inflow $MacroSE_S$ ($ic = 1$)	%
iMacroSE$_b$	Inflow $MacroSE_b$ ($ic = 1$)	%
iMesoSE	Inflow $MesoSE_S$ ($ic = 1$)	%
iMesoSE$_b$	Inflow $MesoSE_b$ ($ic = 1$)	%

(continued)

(continued)

Symbol	Description	Unit
iMicroSE	$= \text{iMicroSE}_S = \text{MicroSE}$	%
indicator of interest	Amount of water allocated to a member in a period of time	L3; L3/L2
iSE	$= \text{iSE}_S = \text{iSefficiency} = \text{Inflow Sefficiency} (ic = 1)$	%
iSE_b	Inflow bSefficiency $(ic = 1)$	%
M1	Water allocated to the most advantaged member (first member)	L3; L3/L2
M2	Water (re)allocated to the least advantaged member (second member)	L3; L3/L2
Macro	Macro level water management	–
MacroSE_S	One of the three levels of Sefficiency	%
MacroSE	$= \text{MacroSE}_S = \text{Macro Sefficiency}$	%
MacroSE_b	Macro bSefficiency	%
Member	Examples: stakeholders, groups, and zones (see M1 and M2)	L3; L3/L2
Meso	Meso level water management	–
MesoSE_S	One of the three levels of Sefficiency	%
MesoSE	$= \text{MesoSE}_S = \text{Meso Sefficiency}$	%
MesoSE_b	Meso bSefficiency	%
Micro	Micro level water management	–
MicroSE_S	One of the three levels of Sefficiency	%
MicroSE	$= \text{MicroSE}_S = \text{Micro Sefficiency}$	%
MicroSE_b	Micro bSefficiency	%
Mn	M1 or M2	L3; L3/L2
n	1 or 2 in Mn, i.e, M1 or M2	–
nb	Non-beneficial index	–
nc	Non-consumptive index	–
NEW	New water supply, $0 \leq \text{NEW} \leq (\text{SW1} + \text{SW2})$	L3; L3/L2
nq	Pollution index	–
NR	Non-Reusable (C–ET = non-ET consumptive water)	L3; L3/L2
ns	Non-useful index	–
O	Outflow (= C + R)	L3; L3/L2
OS	Water from Other Sources	L3; L3/L2
PP	Total Precipitation	L3; L3/L2
PtI	Policy type I	–
PtII	Policy type II	–
PtIII	Policy type III	–
PtIV	Policy type IV	–
q	Quality index	–
R	Returns, return flows/volumes, non-consumptive flows/volumes (= V2 + RP)	L3; L3/L2
R1	Return fraction (= R/I)	–
RE	Resiliency	%

(continued)

(continued)

Symbol	Description	Unit
Reaf	Reallocation fraction = fraction of ZW1 that should be reallocated to M2 (or not be abstracted), $0 \leq$ Reaf ≤ 1	–
RF	Return Flow (return to the main source)	L3; L3/L2
RP	Potential Return (does not return to the main source)	L3; L3/L2
s	Useful; Usefulness index	–
SE	$= SE_S =$ Sefficiency	%
SE_b	bSefficiency	%
Sefficiency	Sustainable efficiency	%
Sequity	Sustainable equity	
Sg	Segment, e.g., Sg14 = segment with M2 in row 1, and M1 in column 4	–
SW	Water Shortage (SW $\geq$ 0) for rows 1 and 2 = The amount of water needed to reach target	L3; L3/L2
SWn	SW of Mn from its Tg = Tgn - Mn	L3; L3/L2
Tgn	Target (Tg) of M1 or M2	L3; L3/L2
TUF_d	Total Unrecoverable Flow along d	L3; L3/L2
V1	Volume of water at section 1 (VU or VA)	L3; L3/L2
V2	Volume of water at section 2 (VD or RF)	L3; L3/L2
VA	Abstracted/Applied water from the main source	L3; L3/L2
VD	Volume of water Downstream after RF in the main source	L3; L3/L2
VU	Volume of water Upstream before abstraction in the main source	L3; L3/L2
W	Weight	–
WaP	Water Productivity	Various
W_{bX}	Beneficial weight of an WPI = X	–
WC1	Desirable Consumption fraction	–
WL	Water Loss	L3; L3/L2
WPI	Water Path Instance	L3; L3/L2
WPT	Water Path Type	L3; L3/L2
W_{qX}	Quality weight of an WPI = X	–
WR1	Desirable Return fraction	–
W_{sX}	Usefulness criterion of an WPI = X	–
WUS	Water Use System	–
X	= WPI for ease of use in equations	L3; L3/L2
X_S	Useful part of X	L3; L3/L2
ZW	Water excess (ZW $\geq$ 0) = The amount of water in excess of target	L3; L3/L2
ZWn	ZW of Mn from its Tg = Mn - Tgn	L3; L3/L2

Chapter 1
Introduction

Life today is in a vast and deep transition. All the major drivers of human civilization, such as economics, governance, science, technology, culture and religion are changing with unprecedented speed. Nature in relative equilibrium with humans for millenniums is now in the middle of this foundational and rapidly shifting transition. Negative natural symptoms are evident in our daily lives and none can reach deeper into our sense of security than water, because "water is life"!

Sever water problems, due to scarcity and/or pollution, are expanding and becoming real for an increasing number of peoples and regions of the world. Most of the population of the world are affected and alarmed, such as India, Middle East, USA (e.g., California), China, Bolivia, Nigeria, South Africa, Iberian Peninsula, just to name a few. Besides the growing focus of the nations on water, its increasing centrality is also evident by many private organisations, big industries, and the actions of international organisations, such as World Bank, WHO, WMO, FAO, UNESCO and OECD. The United Nations Sustainable Development Goals (UN-SDG 2018) have Goal number 6 (Clean Water and Sanitation) with most of the other 16 Goals, such as Goal 2 (Zero Hunger) and Goal 3 (Good Health and Well-Being), being dependent directly or indirectly on Goal 6. We are also living in the "International Decade for Action on Water for Sustainable Development, 2018–2028" as declared by the General Assembly of the United Nations.

Until recent years, almost all that is written and done on water management and planning are from the viewpoints different from water. Generally, one or two other priorities in domains, such as, food, land, economics, health and ecosystem had the centre stage and water was just an input among many others. Historically, such a direction in analyses made sense due to two facts: water was relatively abundant and unrestricted in comparison to other things, and water pollution was much less alarming. This book tries to reverse this tendency, put water as the priority, and present it as the central issue in development. In doing so, the author has given much

© The Editor(s) (if applicable) and The Author(s), under exclusive license
to Springer Nature Singapore Pte Ltd. 2021
N. Haie, *Transparent Water Management Theory*,
Water Resources Development and Management,
https://doi.org/10.1007/978-981-15-6284-6_1

attention into logically, clearly and comprehensively defining the relevant words used in this book. This way, a reader can pursue the ideas and examples without ambiguity, and, hopefully, with less criticism. To continue, let us look at the word 'management' in the title of the book, and briefly explain its relationship with two other important words, viz.: security and governance.

1.1 Water Security, Governance or Management

There are strong associations between these three concepts. Water *security* is relatively new and has a working definition "as the capacity of a population to safeguard sustainable access to adequate quantities of acceptable quality water for sustaining livelihoods, human well-being, and socio-economic development, for ensuring protection against water-borne pollution and water-related disasters, and for preserving ecosystems in a climate of peace and political stability" (UN-Water 2013). This and other water-related definitions and procedures are about three fundamental Pillars, vis.: water quantity, water quality and water benefits, i.e., having adequate quantities of water with suitable qualities to promote livelihood, health, economic and peace. This last item—peace - probably is the distinguishing feature between security and the other two words, and is largely depend on the social contract theory touched in Chap. 5. Anyhow, the concept of three Pillars is significant for the progress of the ideas of this book as presented in Fig. 1.1 for better visualization. Furthermore, "sustainable access" in the above definition is to some extent about equity that we will cover in Chap. 5.

Another word is about *management* of the three water Pillars, which is the focus of this book. To manage is "to work upon or try to alter for a purpose" within the process of conducting or supervising (Merriam-Webster Dictionary 2018) a water system. The alteration process (e.g., allocation, protection, development) is mostly

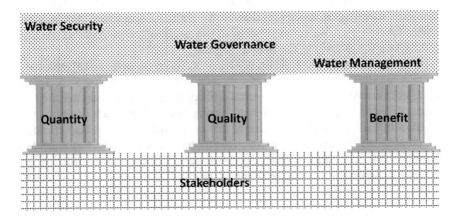

Fig. 1.1 The three pillars of water management

about learning the complex nature of water systems in space-time continuum. It is important to keep the focus on water and its purpose while changing things, i.e., being water centric. We should know that although a water system had mostly one purpose, nowadays, they should have at least two purposes or objectives, such as irrigation and groundwater recharge. In other words, we should understand that management is about output in relation to input with learning and capacity development in its centre. Transparency should primarily relate to capacity building of both water suppliers and users.

Governance is an issue many people, with proper water background or not, with proper governance background or not, feel entitled to talk and to write. The greatest problem is the lack of a generally accepted definition of governance hence it can easily get confused with management or be anything that anyone wants (Tortajada 2010). However, the word governance, in general, has some specific meaning that can be applied in the domain of water. For example, "In a broad sense, governance is about the culture and institutional environment in which citizens and stakeholders interact among themselves and participate in public affairs. It is more than the organs of the government" (UNESCO-IBE 2019). In this regard, Francis Fukuyama explains that this type of definition of governance is primarily about state capacity that is weak and corrupt, which promotes (water) poverty: "Interest in this topic was driven by awareness that global poverty was rooted in corruption and weak state capacity" (Fukuyama 2016). Therefore, it may be confusing if water governance is about water delivery, allocation, efficiency, etc. So, let us mention few brief reflections on the concept of good governance in general, which can help in advancing good water governance. In doing so, our intention is also to sharpen the distinctions between management and governance, while seemingly presenting some interesting ideas. Hence, it is useful to consider four approaches to evaluate the quality of governance, namely, procedural measures, capacity measures, output measures, and autonomy measures (Fukuyama 2013). In analysing these approaches, Fukuyama rejects output approaches and proposes the combination of capacity and autonomy measures as the best apparent solution. The following summary explanations are from the 2013 Fukuyama's paper, unless stated otherwise.

On procedural measures, he discusses the inadequacies of the 10 conditions of the Weberian bureaucracy and states "The idea of bureaucratic autonomy–the notion that bureaucrats themselves can shape goals and define tasks independently of the wishes of the principals–is not possible under the Weberian definition." Then goes on to add "Nonetheless, certain procedural measures would remain at the core of any measure of quality of governance. One would want to know whether bureaucrats are recruited and promoted on the basis of merit or political patronage, what level of technical expertise they are required to possess, and the overall level of formality in bureaucratic procedure."

Capacity measures, according to Fukuyama, include "both resources and the degree of professionalization of bureaucratic staff." He questions possible proxies for aggregate capacity, and differences between potential and actual capacity. Even without corruption, he goes on to state that "a bureaucrat is selected on the basis of 'merit' does not define merit, nor does it explain whether the official's skills will be renewed in light of changing conditions or technology."

Fukuyama excludes output measures as an approach for evaluating governance by presenting three decisive drawbacks. He goes on to affirm that "In fact, it might be better to leave output as an independent variable to be explained by state quality, rather than being a measure of capacity in itself. If output is not a valid measure of state quality, it implies that we also cannot generate useful measures of government efficiency as a measure of state quality, since the latter represents a ratio of state inputs to outcomes." This book is doing this by defining efficiencies as part of the management (not governance) of a multi-level water system. In such a context, it is relevant to mention that activities in water governance based on outputs, such as the OECD Principles on Water Governance (OECD 2015) to mention one important and carefully developed example should be reformulated.

"Autonomy properly speaking refers to the manner in which the political principal issues mandates to the bureaucrats who act as its agent... The fewer and more general the mandates, the greater autonomy the bureaucracy possesses... Conversely, a non-autonomous or subordinated bureaucracy is micromanaged by the principal, which establishes detailed rules that the agent must follow" (Fukuyama 2013).

Figure 1.2 demonstrates differing autonomies (the inverted U curves) for four examples of varying capacity levels. The inverted U shape is because every government bureaucracy can have too little or too much autonomy. However, the lower capacity examples have their optimal point moved to the left, and the higher capacity ones to the right.

Figure 1.3 shows the total space that agencies, countries, etc., can be located. The line sloping downwards and to the left is the line in Fig. 1.2, representing optimal autonomies for varying capacities. "Bureaucracies that were above the line would be hobbled by excessive rules; those below it with excessive discretion. For the past decade, international donors have been advising developing countries to decrease the amount of discretion in the behavior of their bureaucracies." From Fig. 1.3 "it would

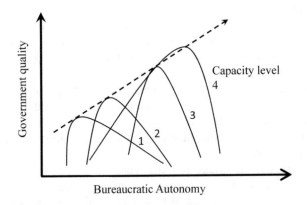

Fig. 1.2 Autonomy, capacity and Government quality relations (Fukuyama 2013)

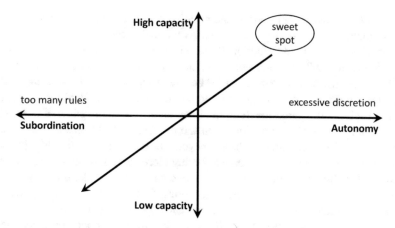

Fig. 1.3 Autonomy and capacity space (Fukuyama 2013)

appear that this is only contingently good advice; in a high capacity state, one would like to have more rather than less discretion" (Fukuyama 2013). This can also apply to different sectors or agencies within a country. He gives the example of India that some agencies need stricter rules, while others, such as "the Hyderabad Municipal Water Authority needs to be relieved of its multiple and conflicting political mandates if it is to function properly."

1.2 Bounded Rationality

In order to develop water management policies, it is helpful to look at bounded rationality instead of perfect or global rationality. The term bounded rationality was introduced and explained by Herbert Simon: "Broadly stated, the task is to replace the global rationality of economic man with the kind of rational behavior that is compatible with the access to information and the computational capacities that are actually possessed by organisms, including man, in the kinds of environments in which such organisms exist" (Simon 1955). In other words, this profound and true way of making decisions states that "access to information" (input), "computational capacities" (including human beings), and existing environment (situation) have indeed degrees, and it is their complex interactions that should lead to rational and impartial decisions. Hence, by definition, proper decision-making is bounded, and depends on the conditions of these three components, which, incidentally, are also used by robots, autonomous driving systems and other Artificial Intelligence applications.

In such a context, only a learning approach that tries to be inclusive (from local to global, indigenous to experts, men and women, traditional and modern…) can bring about a reasonable or good-enough combination of the three components. To explain further, let us consider only one of the three components that influences decisions, i.e., computational capacities of individuals and institutions:

"Simon's focus on computationally efficient methods that yield solutions that are… "good enough", in Simon's terms, involves search procedures, stopping criteria, and how information is integrated in the course of making a decision. Simon offers several examples to motivate inquiry into computationally efficient methods. Here is one. Applying the game-theoretic minimax algorithm to the game of chess calls for evaluating more chess positions than the number of molecules in the universe (Simon 1957). Yet if the game of chess is beyond the reach of exact computation, why should we expect everyday problems to be any more tractable? Simon's question is to explain how human beings manage to solve complicated problems in an uncertain world given their meager resources. Answering Simon's question, as opposed to applying Friedman's method to fit a constrained optimization model to observed behavior, is to demand a model with better predictive power concerning boundedly rational judgment and decision making" (Wheeler 2018).

Let us very briefly consider two of the important points presented in the above quote, namely, optimization and judgment, in relation to water management. Regarding the former, Loucks and van Beek (2005) write: "Hence at best a mathematical model of a complex water resources system is only an approximate description of the real system. The optimal solution of any model is optimal only with respect to the particular model, not necessarily with respect to the real system. It is important to realize this limited meaning of the word 'optimal,' a term commonly used by water resources and other systems analysts, planners and engineers." They go on to state that "Even when combined with efficient techniques for selecting the values of each decision variable, an enormous computational effort may still lead to a solution that is still far from the best possible." In this way, and according to Simon, there should be a clear effort in abandoning the word 'optimum' that means "the best possible" solution, and go toward employing good-enough or reasonable words and approaches.

The second crucial point is judgement in decision-making. Regarding the Berlin Rules on Water Resources, Dellapenna (2008) states that "in English, at least, the word "judgment" carries a connotation that the result is not dictated in any immediate sense by the factual and other inputs that the judge relies upon in exercising judgment." For a water manager, this is the case because s/he has to consider all types of inputs, sometimes non-quantitative or very hard to quantify, such as political and other societal sensitivities. In Chap. 4 regarding sustainable performance of water systems, we will expand two notions linked to judgment (a) Loucks (2002) defines sustainability and affirms that its advancement depends on "Public value judgments", and consequently significant stakeholder involvement, (b) noise or chance variability of judgments is crucial in advancing good-enough water management.

In this view, water policy makers act as satisficers (using Simon's ideas) trying to have good-enough outcomes rather than an optimal one. Loucks and van Beek (2005) write that because of the high complexity involved in water resources management, the idea may prove to be more sustainable if one uses the satisficing decision making. The theory presented in this book makes judgments of water managers much bounded toward sustainability. For example, it will be shown that all the variables and parameters of water management (and they are many) will be logically and reasonably reduced to very few performance indicators for decision making in connection with stakeholders in a specific environment of learning.

References

Dellapenna JW (2008) The Berlin rules on water resources: a new paradigm for international water law. Montpellier, France, International Water Resources Association (IWRA). Available at: https://iwra.org/member/congress/resource/abs568_article.pdf. Accessed 15 Aug 2019

Fukuyama F (2013) What is governance? Center for Global Development, Working Paper 314, Washington, DC

Fukuyama F (2016) Governance: what do we know, and how do we know it? Ann Rev Political Sci 19:89–105

Loucks D, Van Beek E (2005) Water resources systems planning and management: an introduction to methods, models and applications. UNESCO, Paris

Loucks DP (2002) Quantifying system sustainability using multiple risk criteria. s.l., Cambridge University Press

Merriam-Webster Dictionary (2018) Manage. [Online]. Available at: https://www.merriam-webster.com/dictionary/manage#h1. Accessed 3 Mar 2018

OECD (2015) OECD principles on water governance, Organisation for Economic Co-operation and Development. [Online]. Available at: http://www.oecd.org/cfe/regional-policy/OECD-Principles-on-Water-Governance-brochure.pdf. Accessed 7 Mar 2018

Simon HA (1955) A behavioral model of rational choice. Q J Econ 69(1):99–118

Tortajada C (2010) Water governance: some critical issues. Int J Water Resour Dev 26(2):297–307

UNESCO-IBE (2019) International Bureau of Education. [Online]. Available at: http://www.ibe.unesco.org/en/geqaf/technical-notes/concept-governance. Accessed 29 Nov 2019

UN-SDG (2018) Sustainable development goals. [Online]. Available at: http://www.un.org/sustainabledevelopment/sustainable-development-goals/. Accessed 26 Feb 2018

UN-Water (2013) Water security & the global water agenda: a UN-water analytical brief. United Nations University, Canada

Wheeler G (2018) Bounded rationality. [Online]. Available at: https://plato.stanford.edu/archives/win2018/entries/bounded-rationality. Accessed 24 Mar 2019

Chapter 2
Terminology

As will be seen, the theory of this book is vital for water policy, planning and practice due to the high complexity inherent in water use systems, which, in turn, demands a coherent and clearly defined terminology. In this chapter, we define most of the basic vocabularies needed in this book. In doing so, we employed current water terminology from various sectors unified through their formal definitions from valid references, including dictionaries.

2.1 Water Use System (WUS)

"Making water available for its many uses and users requires tools and institutions to transform it from a natural resource to one providing services" (UN-HLPW 2017). Hence, it is crucial to distinguish between resource and service, but most of the time the boundary between them is not sharp enough perpetuating flawed decision making. To overcome this foundational problem, this book *only* focusses on water services and starts with a generic service provider, which we call a Water Use System (WUS) as depicted in Fig. 2.1.

The rectangle in Fig. 2.1 represents the system that is under analysis (WUS), such as a city, industry, crop land, basin, country or countries. The arrows show the nine fixed Water Path Types (WPTs) as given in Table 2.1. Here *fixed* means that all the flows of any WUS has these nine types and symbols, which do not change throughout this book and cases. In the examples and the tables and the figures of this book, if the units of measurement are not important in conveying a concept, they will not be specified. For example, the units of measurement in Table 2.1 are irrelevant as they can be length (L), volume per unit of time (L^3T^{-1}), volume (L^3), or the like and consequently are not specified. Obviously, in any particular situation, all the flows

© The Editor(s) (if applicable) and The Author(s), under exclusive license
to Springer Nature Singapore Pte Ltd. 2021
N. Haie, *Transparent Water Management Theory*,
Water Resources Development and Management,
https://doi.org/10.1007/978-981-15-6284-6_2

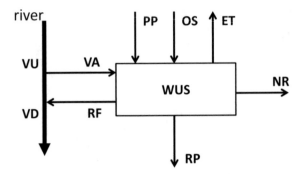

Fig. 2.1 Typical schematic of a WUS (Water Use System) with fixed WPTs (Water Path Types)

Table 2.1 Fixed water path types (WPTs) of a water use system (WUS)

WPT	Description
ET	Evapotranspiration
NR	Non-Reusable
OS	Water from Other Sources
PP	Total Precipitation
RF	Return Flow (return to the main source)
RP	Potential Return (does not return to the main source)
VA	Abstracted/Applied water from the main source
VD	Volume of water Downstream after RF in the main source
VU	Volume of water Upstream before abstraction in the main source

must have the same units for the same time interval, e.g., millimetre or cubic meter per year.

Let us make Table 2.1 a bit clearer:

- Main source usually supplies the most amount of water to a WUS. For example, the above schematic has a river as a main source of water. In many parts of the world, ground water is the main source of water.
- ET: a consumptive water, which means outflow that cannot be reused by the same WUS during a time interval (see Sect. 2.4).
- NR: non-ET consumptive water, such as bottled water leaving a WUS.
- OS: any other flow into a WUS, besides the main source (VA) and precipitation (PP). For example, if there are two sources of water for a WUS and the main source is a river, then any extra inflow say from groundwater is OS.
- VU and VD give the influence of the WUS on the main source and hence the basin.
- Considering V1 for the inflow section and V2 for the outflow section, they are:

    - V1 = Volume of water at Sect. 2.1 which is set as VA or VU
    - V2 = Volume of water at Sect. 2.2 which is set as RF or VD

A few points worth mentioning at this stage:

- In general, a WUS is a construct, meaning that it does not correspond to a specific geographic scale and all its water users. For example, a basin may be considered as a WUS however, we may be only interested to look at the urban water users, even though there are various other types of water users within that basin.
- If the WUS goes beyond one institution or government (i.e., transboundary WUS), they should develop an agreement for considering the WUS as a single system for (comprehensive) analysis.
- It should be noted that a flow in a specific time interval, such as a month or a year, is volume and in this book they have been used interchangeably unless stated otherwise. Indeed, for any given application, the time interval for all the WPTs must be the same making flow and volume equivalent.
- One crucial feature of such a fixed WUS structure is to bring *transparency* into the foundation of the water management process.
- For a particular case, each WPT has greater than or equal to zero Water Path Instances (WPIs). For example, in an industrial application, we may decide that there is no ET, hence, zero WPI for the type ET, and a city water supply has more than one WPI for its RP, such as, leakage, green zone irrigation, hydrants, etc.
- This book clearly differentiates between flows (WPIs) and systems (WUSs) and integrates them for comprehensive, transparent and reasonable water management. This difference is significant and plays an important role in the following chapters.

Finally, a note on calculating ET by the famous Penman-Monteith equation (ASCE-EWRI 2005; FAO 1998). It utilizes various climate variables including mean air temperature based on its minimum and maximum (highly significant variables in a warming world). For an analysis of the domains of the mean temperature, please refer to the open access paper by Hai et al. (2018). For a complete analysis of the eight-dimensional domain (or space) of the Penman-Monteith equation, please refer to Haie et al. (2019). My personal webpage at Haie (2020) has a version of this paper with a frequently asked questions file that explains in more details some of the concepts associated with the domain analysis performed in this paper. Let me clarify the notion of the behaviour of the domain of an equation, which is different from sensitivity analysis common in many studies. For this, let us see a simple example by considering $y = x^2 + 1$, which is symmetric along the y-axis, without real zeros, nonlinear, etc. These characterize the behaviour of the domain of y and are of great value to those interested in such an equation. Of course expanding from two to eight dimensions has its difficulties.

2.2 Pillars: Quantity, Quality, Benefits

Figure 1.1 shows the idea of the three fundamental water Pillars, namely water quantity, water quality and water benefits (economic or not). One of the most difficult parts of this book is to continuously remember that the theory presented here is systemic, i.e., based on a predefined system (WUS) that explains all of its water quantities as described in the previous subsection. There are, of course, clear distinctions between the three Pillars of water management. Water quantity is the liquid/vapor part and water quality and benefits are the attributes of water quantity. For example, one litre of water is one litre of water whether it is beneficial or not, polluted or not.

In many studies, the two water attributes manifest their influence on the performance of systems through two weights, varying from 0 (worst) to 1 (best). This means that each WPI of a WUS (denoted as X for ease of reference in symbols and equations) has a beneficial weight (W_{bX}) and a quality weight (W_{qX}) with their product giving Usefulness Criterion (W_{sX}), as defined below:

- Beneficial Weight (W_{bX}) represents the fraction of a water flow that has value for the objectives of a WUS.
- Quality Weight (W_{qX}) represents the fraction of a water flow that has quality for the objectives of a WUS.
- Usefulness Criterion (W_{sX}) represents the fraction of a water flow that has value and quality for the objectives of a WUS.

Although there will be more explanations about these weights and their definitions, for now let us write their equations in Eq. (2.1).

$$W_{bX} = \frac{X_b}{X}, W_{qX} = \frac{X_q}{X}, W_{sX} = W_{bX} * W_{qX} \tag{2.1}$$

Let us see some of the preliminary explanations behind the two attributes:
Beneficial weight:

- In setting this weight, no consideration is given to water qualities (i.e., $W_{qX} = 1$). This is essential because the focus must be on the nature of the WUS and the degree of the value of water in its functioning. The influence of water is captured in the quality weight.
- It is set relative to the objectives of the WUS.
- It covers both the value and the cost as related to a water flow (WPI).
- Effective participation of all the stakeholders is crucial.

Quality weight:

- It is the aggregate of the measurements of all the necessary quality parameters of a WPI (= X). They are generally satisfactory in relation to certain predefined minimum and maximum values.

- It is relative to the required water quality norms for the objectives of the WUS. Sometimes, these norms divide the quality weights into two states, such as, satisfactory/unsatisfactory, suitable/unsuitable, 0.98/0.11.
- W_{qET} is always one, and under some special conditions, such as bottled water that goes out of the WUS, W_{qNR} is also one.

Of course, setting weights is not easy and we will explain them more. For now, let us present some general points regarding the weights:

- In locations with partial or no data, these two weights and quantities of WPIs can be estimated, but with active stakeholder involvement without ignoring most of the above points.
- Both uncertainty and sensitivity of the weights are important in setting them. For example, a lesser sensitive weight can be set with less effort than a higher sensitive one. This means that higher error or uncertainty in the former can still result in better outcomes and more accurate performance than lesser uncertainty in the latter (Loucks and Van Beek 2005).
- A WUS in a water scarce region may get a lower W_{bX} for its inflow WPTs, if its activity is based on high water consumption, such as, water bottling plants or irrigation of water intensive crops.

To include both quality and beneficial attributes and their trade-offs in the performance indicators, we need the concept of Usefulness Criterion (W_{sX}) as defined above and given in Eq. (2.1); but why multiplication and not summation? The reason is that to calculate one single criterion from multiple attributes is through multiplication of the attributes, whereas the presentation of the various (possible) states of a single attribute is through the summation of those states. In other words, the Usefulness Criterion is the product (not the sum) of the quality and beneficial attributes. The multiplication also makes sense thinking as follows: if the benefits of a flow of water is zero, it means that the contribution of that flow to its usefulness is also zero no matter what its quality is. On the other hand, if the water is so polluted that makes the flow totally improper for a particular purpose, then again its contribution to usefulness is zero no matter what its beneficial potential is. Besides associating W_{sX} to WPIs of the WPTs (Table 2.1), Table 2.2 gives their aggregations.

Three points worth mentioning here. First, Table 2.2 gives the total values although the word 'total' is not used. For example, Consumption gives the total consumption of a WUS. Second, the theory in this book is water-centric (analysing from a water perspective) and in this context the 'level' of management (more in Chapter 4) is defined according to the flows considered. This means that the nine fixed WPTs defined in Fig. 2.1 are not present at the three levels utilized in this book, viz.: macro, meso and micro as follows:

- Macro does not consider VA and RF. It considers macro-level I, C and R.
- Meso does not consider VU and VD. It considers meso-level I, C and R.
- Micro does not consider VU, VD and R. It considers micro-level I and C.

Table 2.2 Aggregating flows and applying Usefulness Criterion

Symbol	Expression	Terminology
I	$V1 + OS + PP$	Inflow
R	$V2 + RP$	Return
C	$ET + NR$	Consumption
O	$C + R$	Outflow
UI	I_S	Useful inflow
UR	R_S	Useful return
UC	C_S	Useful consumption
UO	O_S	Useful Outflow

Third, for each WUS, the law of mass conservation or water balance must be satisfied (Sutcliffe 2004). It states that total inflow into a system is equal to total outflow plus change in storage within any given time interval, and as such multilevel management of a WUS necessitates water balance at all levels. However, the micro level analysis is not based on water balance because it does not consider the returns, and as such, it is prone to errors. It is included in this book because most of the current analyses in papers, reports and projects are done using some form of micro level analysis. Furthermore, it should be mentioned at this point that the change in storage is considered to be zero as will be explained fully in the Chapter on the Theory (see Sect. 3.1). Consequently, the water balance in this book has the form of Eq. (2.2) using the flows presented in Table 2.2:

$$I = O = C + R \tag{2.2}$$

A caution needs to be exercised in applying weights to the sum or difference of the flows. Remembering that a WPI ($= X$) is a real flow that can be measured with its own two weights, let us examine two distinct combination properties:

Property 1: assume that a farm (WUS) has three WPIs that obey the water balance as follows:

$$X_1 = X_2 + X_3 (\text{e.g., } VA_f = ET_f + RF_f, \text{ f for farm})$$

Applying the corresponding weights, say the beneficial ones, and considering Eq. (2.1) we get:

$$X_{b1} \neq (X_2 + X_3)_b \rightarrow W_{bX1} * X_1 \neq W_{bX2} * X_2 + W_{bX3} * X_3$$

The reason for the inequality, of course, is the fact that the three weights are not equal having in mind that X_{b1} is the beneficial fraction of X1.

Property 2: Let us consider the following:

$$Y = X_1 + X_2 \rightarrow Y_b = (X_1 + X_2)_b = X_{b1} + X_{b2}$$

Here, Y is a sort of a placeholder, an aggregate 'flow' consisting of two or more real flows (X), and does not exist in nature. Another way of maintaining equality depends on the weight of Y (W_{bY}). It can be shown that for this to happen, the weight of the aggregate flow must be computed by Eq. (2.3), which uses a simple weighted averaging:

$$W_{dY} = \frac{\sum_{j=1}^{J} X_{dj}}{\sum_{j=1}^{J} X_j}, d = b, q, s \tag{2.3}$$

X_j is an element of an aggregate flow Y, J is the number of elements in Y, and d is an index for the desirables, i.e., along beneficial, quality or useful dimensions (see below in Sect. 2.4). For example, to find the beneficial weight of Inflow (W_{bI}), we have d = b, Y = I, J = 3 (e.g., X_1 = V1, X_2 = OSa, X_3 = OSb, with PP = 0), and X_{bj} has three elements, the first one being X_{b1} meaning the beneficial part of V1.

2.3 Water Use and Reuse

The word 'use' is defined as "to put something into your service for a purpose" (Cambridge Dictionary 2018). Hence, water use is defined as the water applied to a WUS for its objectives during its time interval. Regarding this quantity definition, three points worth mentioning:

- A WUS (Fig. 2.1) should have more than one objective, which can be water supply to populations, crops, industries, and aquifers, i.e. groundwater recharge (broadly securing a healthy downstream).
- In general, not all of the applied water to a WUS goes toward fulfilling its objectives (e.g. irrigation water also goes to weeds). In other words, water quantity is different from the degree that it is beneficial (W_{bX}) in fulfilling its objectives.
- This book deals with WUSs and not water resources (see Sect. 2.1), and explicitly distinguishes the three Pillars (Sect. 2.2). This means that water use, access, available, applied and allocated are all the same because our focus is a WUS (Water Use System). For example, under the following situations, we should conclude that there is no WUS:

  - If not allocated cannot be used
  - If not available cannot be used
  - If not accessible cannot be used.

However, this does not exclude the complexities associated with having accessible water. As mentioned here, the first step is a robust and transparent methodology to distribute water equitably among various members. For example, UN (UN-Water 2019) states that water must be accessible within 1000 m and 30 min. Consequently, if the water management of a region adopts these thresholds, but does not implement them, then water allocation, availability, access and use are zero for that particular WUS or a member within it.

Reused water is the amount of Outflow of a WUS that becomes part of its Inflow during its time interval. The amount of Outflow that is reusable is also called non-consumptive water, which consists of all the return flow (R = V2+RP). The actual amount of water reused varies from place to place and depends on economy, technology, culture, water resources, etc. If there is reuse, it shows up as an OS with its own beneficial and quality weights. These have lower numerical values because reusing water has costs, and are generally more polluted. What happens to the water that is the difference between all the return flow (reusable) and reused water is of no interest to the WUS unless it becomes one of its objectives. For example, sometimes this difference is beneficial due to legal limitations. A country (WUS) may be obliged by treaties to deliver a minimum amount of water with a minimum quality to a downstream country. These two minimums should affect the management and design of the activities in the upstream country, which, of course, cannot reuse all its return flows.

2.4 Binary Opposites in Water

A small list of opposite words most commonly seen in water literature is given in Table 2.2 with some comments as related to WUSs in general.

A few notes regarding Table 2.3:

- Among many ways to understand beneficial flow is the document of the Bellagio Principles that states "Valuing water means recognizing and considering all the benefits provided by water that encompass economic, social and ecological dimensions" (UN-HLPW 2017). The important fact about this statement is that benefits are not exclusively about economic calculations, but also about all sorts of values received by using water.
- Beneficial, Quality and Useful flows correspond to beneficial weight, quality weight and their combination usefulness criterion, respectively. In a compact form, they are called the desirable flows with the index 'd', which includes beneficial (b), quality (q) or useful (s).
- Water consumption has two parts: ET and what is called non-reusable outflows (NR in Table 2.1). ET is separated from other non-reusable outflows because of its special importance in agriculture, hydrology and ecosystems.

  - Many times the word 'consumption' is utilized as water 'use', which confuses the consumptive and non-consumptive nature of water. Water has two states in

Table 2.3 Binary terminologies used in water

Word	Opposite	Definitions and Comments on Word (1st column)
Available	Unavailable	Same as water use
Beneficial	Non-beneficial	Fraction of a water flow that has value for the objectives of a WUS
Consumptive	Non-consumptive	Outflow that cannot be reused by the same WUS during a time interval
Desirable	Undesirable	Compact form of referring to beneficial flow, quality flow or useful flow explained in this table
Quality	Polluted (non-quality)	Fraction of a water flow that has quality for the objectives of a WUS
Reusable	Non-reusable	Return flow = V2+RP, (Non-reusable = NR)
Satisfactory	Unsatisfactory	Falls within the allowed limits defined by the thresholds (minimum and maximum values) of an attribute
Useful	Non-useful	Fraction of a water flow that has value and quality for the objectives of a WUS

contrast to other resources, which Chap. 3 make them clearer. I have chosen to utilize the word 'consumption' as defined in Table 2.3 but some experts might see this choice frustrating. In truth, the word does not matter because concepts are crucial and we have been careful throughout this book in accurately defining the words we utilize. Hence, instead of consumptive and non-consumptive, the reader can adopt his/her own words to reveal explicitly the two states of water; but they must not convey overlaps with the other terms given in this book.

- Quality flow is generally determined by Satisfactory (Table 2.3) attributes and parameters.
- The word 'fraction' means a physical quantity of water and/or an equivalent amount (in percentage). Examples:

  - Quality flow is an equivalent amount. For example, a four cubic-metre tank of polluted water cannot be divided physically into three cubic-metre of quality water and one cubic-metre of only polluted water.
  - Beneficial flows may contain an equivalent amount for less tangible social values, such as recreational values.
  - Non-beneficial flows like weed ET can be estimated as an amount of real water.

- The phrase "quality water" in the UN definition of water security (UN-Water 2013) is used for each WPI to refer to an equivalent "quality water" or quality flow. For example, public water supply that is treated to achieve proper quality for human drinking has the best quality ($W_{bX} = 1$).
- The idea of thresholds in the above definition of Satisfactory is central to many studies, including some of those presented in this book (Loucks and Van Beek 2005). Almost everything in the socioeconomic and natural systems function within minimum and maximum values (thresholds), which are vital to their

sustainability. This means that one of the initial major policy decisions for a WUS is to define its quantity and quality thresholds involving all the stakeholders in a learning mode as we repeat throughout this book. This is significant, particularly under water scarcity, because delimits the types of benefits (activities and investments) that are needed or possible.

2.5 Water Loss of Flows and Systems

In literature, Water Loss (WL) is a non-beneficial quantity that may not be reused toward the objectives of the WUS during a time interval. Hence, it is of great interest to reduce them as much as possible. According to this definition, WL is a water quantity along the beneficial Pillar and consists of two components: non-beneficial consumption and non-beneficial return, which add up to non-beneficial Outflow ($C_{nb} + R_{nb} = O_{nb}$).

However, many words and phrases in this book are getting a wider and deeper meaning, and WL is not exempt. For example, consider the following two ways that have influence over our WL developments. First, we integrate the three Pillars of water management in order to come up with a comprehensive analysis, which is systemic (WUS) with fixed WPTs, hence giving WL other dimensions such as pollution that may be important to some sectors of water use. Second, at the system level, the improvements consider the differentials along the three Pillars, which further advances the reality of water use. For example, water balance (Eq. 2.2) represents a zero differential along the quantity Pillar.

In order to progress toward a wider meaning of WL for each flow, we should remember that weights along the two Pillars, namely, water benefit and water quality, divide each WPI into two parts, viz.: desirable and undesirable depending on the objectives of a WUS. This means that for the same flow desirable weight + undesirable weight = 1. Weights give the fraction of the flow that is desirable relative to a specific Pillar and are called beneficial flow, quality flow and consequently useful flow (Table 2.3) in accordance with the objectives of a WUS. The remaining fraction of a flow is undesirable and called non-beneficial flow, non-quality (polluted) flow and non-useful flow. In other words, the two attributes of every WPI are divided into beneficial/non-beneficial (X_b/X_{nb}), and quality water/polluted water (X_q/X_{nq}). Utilizing these partitions, Usefulness Criterion (weight) produces useful/non-useful flows (X_s/X_{ns}). As such, WL of a flow along a Pillar is its undesirable, e.g., the non-beneficial applied water (VA_{nb}), ET_{nb} or the pollution level of the return flow (R_{nq}).

However, knowing the WL of each flow (WPI) raises the question of how to calculate the WL of a WUS, i.e., the integrated inflows and outflows altogether. For a system (WUS) the desirable and undesirable must be defined in terms of the differentials between inflow and outflow, because the undesirables of the outflows depend on the degrees of the desirables of the inflows. Hence, the WL of a WUS must consider all of its flows using water balance:

$$I-O = 0 \rightarrow I_d + I_{nd} - (O_d + O_{nd}) = 0 \rightarrow I_d - O_d = O_{nd} - I_{nd}$$

$$WL_d = I_d - O_d, 0 \le WL_d \le I_d, d = b,q,s \qquad (2.4)$$

The subscript 'd' is a compact form of the desirables along two Pillars, i.e., beneficial (b), quality (q) and, consequently, useful flows (s). The subscript 'np' gives non-beneficial, non-quality and non-useful fractions, having in mind that, in general, the desirable weights (W_d) of the inflow WPIs have influence on W_{nd} of the outflow WPIs. WL_d, Eq. (2.4), is the undesirable of a WUS along 'd', and is equal to desirable Inflow (I_d) minus desirable Outflow (O_d), or undesirable Outflow minus undesirable Inflow for the entire system. For example, the quality WL (= WL_q = WL along the water quality Pillar) of an industry is the difference between its quality Inflow (= I_q) and quality Outflow (O_q), or its Outflow pollution (= O_{nq}) minus its Inflow pollution (O_{nq}). The equalities put on the conditions of Eq. (2.4) are very rare or rather non-existent. Additionally, flow WL is rarely used because it is more common to employ the adjectives non-beneficial, polluted or non-useful. Let us become clearer about the three differentials presented in Eq. (2.4):

- WL_b (WL along the beneficial Pillar) is due to non-beneficial quantities excluding water quality ($W_{qX} = 1$). WL_b is the total beneficial inflow into a WUS minus total beneficial outflow from the same system. To calculate WL_b, according to Eq. (2.4) we need all the WPI quantities of water (three inflow plus four outflow types) and their corresponding beneficial weight s (W_{bX}).
- WL_q (WL along the quality Pillar) is due to pollution excluding water benefits ($W_{bX} = 1$). Water quality is a resource and pollution reduces its availability. WL_q is the total quality inflow into a WUS minus total quality outflow from the same system. This kind of WL does not correspond to a real quantity of water but rather an equivalent amount. One important question is whether WL_q should be considered as a cost to a WUS or as an uncompensated cost to downstream users.
- WL_s (WL along the beneficial and quality Pillars, or along usefulness) is due to non-useful water. It is the total useful inflow into a WUS minus total useful outflow from the same system. It should be remembered that usefulness depends on the degree of beneficial and quality attributes of various WPIs of a WUS.

Besides this fundamental understanding, Water Loss (sometimes called water waste) has been used under other contexts, which raise questions, such as:

A. What is the relationship between WL and the least advantaged population?
B. What is the relationship between WL and the number of objectives of a WUS?
C. Does it integrate all the three Pillars of water management?

Globally the most important issue has to do with the use of water by the least advantaged farmers and urban population. They divert water without permission for their livelihood, which in reality is not a WL because nobody can declare that the use of water by such members of our society is undesirable, so appropriately is counted as part of the non-revenue water. This is one of the reasons that satisfying the water needs of the least advantaged must be explicitly one of the objectives of a WUS and actively pursued to be solved (see Chap. 5 on equity). Of course, there are other situations that water is diverted illegally, which presents as one of the reasons for serious implementation of smart systems (see Sect. 3.3).

Two famous undesirables are leaks from water supply systems, and return flows from irrigated lands. In general, there should be a determined effort by water managers to reduce them, particularly in water scarce regions. However, it is impossible to reduce them to zero, meaning that their beneficial weights mostly should not be zero, because they contribute to various water bodies. In other words, not all of them is Water Loss, because at each moment, it should be assumed that the water managers are doing their best to minimize them.

Sometimes W_b and/or W_q make the usefulness of a WPI low, which can give the idea that ignoring them altogether is good enough. For example, considering all the pipe leaks non-beneficial, may lead to set their W_b to zero. However, there are two reasons that should motivate us to avoid such a common temptation. First, in-line with what just mentioned, sometimes there are WPIs that their usefulness is not low, and therefore excluding them as WL from some considerations is rather premature. Second, sometimes due to trade-offs of the three Pillars, a small value can lead to informative influences on the performance of systems as explained in Chap. 4.

Although the nature of all these parts of WL are different, they all convey the various ways that a WUS makes water undesirable as can be seem from Eq. (2.4). This is highly important in water scarce regions, having in mind that with lower water quantity, the impact of pollution is higher. For example, should water, beverage and other types of companies be allowed to operate in a water scarce region? Should irrigated lands increase? Chapter 4 and Chapter 5 will respond to these and other questions by presenting appropriate approaches.

2.5.1 Unrecoverables

One of the interesting expressions emerged from Eq. (2.4) that is particularly important for the sustainable performance of a WUS, such as in Chap. 4, has the following equation, remembering from Sect. 2.2 that $I_d - R_d = (I - R)_d \neq C_d$:

$$WL_d = I_d - (C + R)_d = I_d - C_d - R_d \rightarrow C_d + WL_d = I_d - R_d$$

This expression is Total Unrecoverable Flow (TUF) along d as given in Eq. (2.5):

$$TUF_d = I_d - R_d = C_d + WL_d = C + R_{nd} - I_{nd}, d = b, q, s \qquad (2.5)$$

Equation (2.5) states that TUF_d is the difference between Inflow and Return along d. It is also the sum of the Consumption and say the differential between non-useful Return and non-useful Inflow. We can also observe from Eq. (2.5) that as WL_d goes toward zero (one of the important aims of water management) TUF goes toward C.

For example, assuming that $W_{bI} = W_{bR} = 1$, then $TUF_b = C$. In other words, for d = b, TUF_b becomes the difference between beneficial Inflow and beneficial Return, which at the limit equals to Consumption. This emphasizes the importance of the notion of differentials that is an integral part of this book.

As to its naming, TUF has a portion of Inflow, Return, and Consumption, which makes it a difficult term to name. For example, Keller and Keller (1995), the first to recognize its importance, probably knew of the difficulty because they called it "effective use" and in the next line "effectively consumed", without defining the word effective. In the original formulation of Sefficiency (Haie and Keller 2012) and subsequent publications, we maintained that terminology (i.e., effective consumption), however later realised that it is composed of consumptive, non-consumptive, desirables, and Water Losses, and cannot be simply called "effective consumption" whatever the meaning of the word effective. Another example is related to the fact that TUF_d may include water quality/pollution (see FIW2, Chapter 3). The existence of such a concept may had been evident to many experts but they had difficulty to find its expression, let alone naming it properly. Within this context, many experts and global organisations merely included pollution into the concept of consumption without considering in their calculations, which created much confusion (Haie and Keller 2014) including institutional fragmentation by 'outsourcing' pollution problems to a completely separate department or entity within a government.

Finally, let us repeat for clarification the differences between desirables, WL and TUF. Desirables are flow (WPI) features along quality, benefit and usefulness, but WL and TUF are system (WUS) features. It is easy to recognise from Eq. (2.5) that TUF_d ($> WL_d$) is a generalised form of WL, because TUF considers differentials, distinguishes between revenue and non-revenue flows, and can include pollution.

References

ASCE-EWRI (2005) The ASCE standardized reference evapotranspiration equation. American Society of Civil Engineers, Reston, VA, USA

Cambridge Dictionary (2018) Use. [Online]. Available at: https://dictionary.cambridge.org/dictionary/english/use#dataset-cacd. Accessed 24 Oct 2018

FAO (1998) Crop evapotranspiration: guidelines for computing crop water requirements. FAO: Food and Agriculture Organization of the United Nations, Rome, Italy

Haie N, Keller A (2012) Macro, meso, and micro-efficiencies in water resources management: a new framework using water balance. J Am Water Resour Assoc (JAWRA) 48(2):235–243

Haie N, Keller A (2014) Macro, Meso, and Micro-efficiencies and terminologies in water resources management: a look at urban and agricultural differences. Water Int 39(1):35–48

Haie N (2020) My Google website. [Online]. Available at: https://sites.google.com/view/naimhaie5. Accessed 13 April 2020

Haie N, Pereira R, Yen H (2018) An introduction to the hyperspace of Hargreaves-Samani Reference evapotranspiration. Sustainability 10(11):1–18

Haie N, Pereira R, Machado G, Shahidian S (2019) An introduction to the hyperspace of Penman-Monteith Reference evapotranspiration. Int J Hydrol Sci Technol 9(1):48–64

Keller A, Keller J (1995) Effective efficiency: a water use efficiency concept for allocating freshwater resources. Winrock International, Arlington, Virginia

Loucks D, Van Beek E (2005) Water resources systems planning and management: an introduction to methods, models and applications. UNESCO, Paris

Sutcliffe JV (2004) Hydrology: a question of balance. IAHS Press, sl

UN-HLPW (2017) Bellagio principles on valuing water. United Nations, Sustainable Development Goals (SDG), High Level Panel on Water. [Online]. Available at: https://sustainabledevelop ment.un.org/content/documents/15591Bellagio_principles_on_valuing_water_final_version_ in_word.pdf. Accessed 29 Sept 2017

UN-Water (2013) Water security & the global water agenda: a UN-water analytical brief. United Nations University, Canada

UN-Water (2019) Water as a global issue. [Online]. Available at: https://www.un.org/en/sections/ issues-depth/water. Accessed 10 Aug 2019

Chapter 3
Theory

The definition of 'theory' is "a system of ideas intended to explain something, especially one based on general principles independent of the thing to be explained" (Oxford Dictionary 2018). This book presents such a theory based on five foundational ideas using independent principles, and a basic law of nature in order to guide water management process toward sustainability. Furthermore, the thought of developing a 'theory' for water use systems emerged when the author was reading the book of the late eminent scholar John Rawls (1999) entitled "A theory of Justice". In this famous book, he advanced a theory based on three fundamental ideas; one of which (the Difference Principle) gave impetus to some of the important developments of this book as will be explained. Crucially, the emergence of a universal aggregative indicator and an objective distributive approach for water use systems are significant outcomes of the theory, which served as the foundation.

Before continuing to the theory itself, let me highlight that this book is an introduction trying to fulfil the prediction of its appearance by many authors, such as:

- "One can reasonably expect that future literature and theory on water allocation will emerge to be usefully coherent rather than contrarian or incomplete" (Lankford 2013).
- "Yet, despite the gravity of the current water crisis—which is projected to become even more severe – our theoretical and analytical models do not adequately explain inequitable water access and distribution, or how equity might be achieved" (Lu et al. 2014).

© The Editor(s) (if applicable) and The Author(s), under exclusive license
to Springer Nature Singapore Pte Ltd. 2021
N. Haie, *Transparent Water Management Theory*,
Water Resources Development and Management,
https://doi.org/10.1007/978-981-15-6284-6_3

3.1 Five FIWs (Foundational Ideas About WUS)

The five Foundational Ideas about WUS (Balance, Consumption, Quality, Benefits, Demand) and their basic sub-ideas are enumerated in Table 3.1 and subsequently briefly explained. These Foundational Ideas that together form the theory of managing water use systems have generic and binding interpretations as presented in the rest of this subsection. Most of the Foundational Ideas are simple and common knowledge, however, putting them together and integrating them for decision making in a water management setting is new and complex.

First FIW: Balance

FIW1a) A Water Use System (WUS) must obey water balance for a specified time interval.

Water balance is an crucial part of water management. Employing any tool (e.g., models and indicators) for the analysis of a water system that is not based on water balance is almost always erroneous. Water balance describes the nature of the distribution of water quantity within a Water Use System (WUS). In other words, it is an accounting mechanism that ensures proper balance and is an important check on possible errors. It is descriptive and does not reveal how a WUS is performing.

Table 3.1 FIWs (Foundational Ideas about WUS) for a water management theory

FIW1 Balance
(a) A Water Use System (WUS) must obey water balance for a specified time interval
(b) Data of the three Pillars should be transparent and reasonable which technology can promote
(c) All the stakeholders of each inflow and outflow must be enabled to effectively participate in its management
FIW2 Consumption
(a) Inflows into a WUS become consumptive and non-consumptive outflows (binary state)
(b) Water consumption is a crucial factor of water scarcity
(c) Both consumption and inflow should be controlled within the context of integrated trade-offs of the three Pillars
FIW3 Quality
(a) There are water quality norms and regulations for different objectives
(b) Downstream water quality should be protected
(c) Water pollution must be treated adequately for people and nature
FIW4 Benefit
(a) Water benefits should reveal multiple values and costs to all the stakeholders
(b) The overriding benefit is adequate water supply to all the population
(c) Investment should benefit multiple water objectives while eliminating allocation extremes
FIW5 Demand
(a) Water demand management has priority over water supply infrastructure
(b) Adaptive water management should advance at multiple levels within a learning approach
(c) Performance of the interactions between water demand and supply depends on the differentials of its three Pillars

Because of its centrality to water management, it is crucial to promote learning approaches that make water balance more robust and transparent. Most of the time, balancing water is complex and needs an explicitly designed learning methodology that properly includes all types of stakeholders that are essential for advancing with better data in time.

Water balance has four parts, viz., total inflow, total outflow, change in storage, and time interval. The first two parts are rather obvious and the third one is explained in the next paragraph. Here, it is sufficient to mention that water balance and its first three parts must be associated with a specific time interval, such as, a day, month or year. A good reference for a better understanding of scales and model time periods is Loucks and van Beek (2005).

The third part, i.e., the change in storage of a WUS, is considered to be negligible for some periods of analysis, such as a day or a year, hence making inflows equal to outflows. If needed, reasonable changes to WPTs, such as OS or RP would establish the needed water balance. A WUS, such as a farm, city, industrial plant or region, is different from a water resource, such as a river or an aquifer. The former does not have change in storage within an appropriate period of time, but the latter may have. To ignore such a fundamental distinction between resource and WUS (see also Sect. 2.1) is a major point of misunderstanding and mismanagement. A resource is basically a storage, however a WUS interacts with water flowing through it with relatively negligible storage meaning that even if there is a change in storage, it is negligible relative to the total inflow into the WUS during its time interval. This specific time depends on the nature of the WUS during which any change in storage should be ignored. Let us see three examples:

- If a farmer pumps groundwater for irrigating his farm (WUS), the water balance of the farm during a season or a year can be written as inflows = outflows even though on a daily or monthly basis this may not be correct. However, during the year, the pumping caused the storage of the resource (groundwater) to diminish.
- In many studies, water supply and wastewater systems (WUS) are routinely considered as having no storage, at least for monthly, seasonal or annual studies. There are, of course, storage reservoirs in almost all the urban areas, however, their combined change in storage is negligible, say on a monthly or yearly basis, and indeed it is almost zero relative to the total flow into the WUS.
- Normally, a dam has almost the same amount of water storage at the beginning and at the end of a chosen time interval, such as a year. This means that its inflows are the same as outflows during that period, even though there may be considerable storage (more inflows than outflows) at one particular period, e.g. January to March.

FIW1b) Data of the three Pillars should be transparent and reasonable which technology can promote.

The data of the three Pillars (water quantity, quality and benefits) effectively used to make final decisions must be transparent, having in mind that the start of good management is data transparency. This means that the data should be open, clear and easy to identify and understand by non-specialists. For example, the data presented should clearly demonstrate inflows, outflows (i.e., all the WPIs) and their balance (FIW1a). The quality and benefits attached to each WPI should be easily recognisable.

Transparency is crucial because the data of the three Pillars are very complex and sometimes not directly measurable. However, lack of measured data must not be the cause of eliminating a particular WPI. The accuracy of the data for each WPI should be reasonable in accordance with a proper combination of measured data, local knowledge, and expert estimate. Sometimes one of these three sources of data is not available and should be promoted. For measurable data, utilizing Information and Communication Technologies (e.g., sensors and intelligent systems) can go a long way to increase the amount of data and hence the accuracy of the final decisions. Local knowledge is important and should be discovered through direct involvement of the stakeholders. Expert estimates should be precautionary and consider comprehensive and integrated solutions to water rather than focussing on other specific aspects such as economic, ecosystem, health, food, land, or construction.

FIW1c) All the stakeholders of each inflow and outflow must be enabled to effectively participate in its management.

Each of the inflows and outflows of a WUS, i.e., all of its WPIs, has various stakeholders. Each stakeholder should choose its WPIs of interest including their quality and benefit, and inform the decision makers. So the phrase "all the stakeholders" in FIWs means the totality of the stakeholders that have shown interest in a particular WPI. The active involvement of stakeholders in decision making must be ensured through a transparent, independent and free process. Institutions of water management should devise an adaptive and autonomous reporting and monitoring of stakeholder involvement. In this process, all the stakeholders should be enabled to participate. Examples of stakeholders are urban, irrigation, industry, river, aquifer, delta, downstream, and other special groups, such as marginalized and (water) poor people. Furthermore, important concerns, such as the precautionary principle and minimization of environmental harm, should be properly handled by ensuring active involvement of stakeholders that defend them. In the final analysis, it is the responsibility of water institutions with the full cooperation of the water companies to secure the active involvement of all the stakeholders. In doing so, fairness, impartiality and transparency will increase, promoting the trust and confidence of the stakeholders in sharing their data truthfully, which makes water management more robust.

Second FIW: Consumption

FIW2a) Inflows into a WUS become consumptive and non-consumptive outflows (binary state).

Inflow into a WUS makes water available for its services, which in turn becomes outflow according to water balance. The amount of the outflow that becomes unavailable to the same WUS for reuse in a time interval is called water consumption. In other words, a fraction of the water applied to any WUS becomes water consumption, such as, crop evapotranspiration (ET), reservoir evaporation, water transfers out of the WUS, evaporation during taking a shower, and water consumed to produce meat and food.

The definition of the word 'consume' in dictionaries (such as Oxford and Cambridge) includes eating and drinking as clear examples of consumption. However, this is not correct when applied to water use. Let us present a candy example taken from these dictionaries. A candy totally disappears after eating it and cannot be reused, i.e., it is completely consumed. However, drinking a litre of water, a fraction of it leaves our bodies as urine and becomes available in the same day and can be reused (treated or not), meaning that the one litre was partially consumed. Conversely, the fraction of the outflow that the same WUS has the possibility to reuse within its time interval is called non-consumptive water. The word 'possibility' in the previous statement is important because in many cases reuse is not only dependent on the existence of water, but also on cultural acceptance, financial resources, and management expertise.

Generally, the total outflow from a WUS has two states, viz.: unavailable and available, which are labelled as consumptive and non-consumptive, respectively. It is of interest to know that such a binary state of water is rare, and other resources do not exhibit this characteristic, or its occurrence is negligible. For example, fuel, electricity and food are only consumed (single state), which justifies the logic of utilizing the word 'consumer'. Filling the tank of a car with say 10 litres of fuel and driving it until the tank is empty, all the 10 litres get consumed and there remains no fuel for reuse by the same car within the time interval (say one day). However, after taking a shower, most of the water can be reused (treated or not), meaning that only some of the water was consumed (e.g., evaporated).

Besides this crucial quantity aspect of consumption, we can also consider two states for the other two Pillars, i.e., water quality and benefits. For example, if a return flow is polluted to the extent that cannot be reused by the same WUS within a period of time, then it should be accounted as unrecoverable, which by definition is a kind of consumption that should influence water management and planning.

FIW2b) Water consumption is a crucial factor of water scarcity.

There are various reasons for water scarcity, such as mismanagement, lack of infrastructures, and physical shortage. One of the overall objectives of the current theory is to substantially reduce the first reason. The second one is mostly because of economic issues, which with the first reason are influenced by corruption. The third one is due to the absence of water resources, which, in general, means low precipitation or its improper pattern. As explained above, water consumption decreases water availability for a WUS, hence is one of the real contributors to the appearance of water scarcity. Indeed, places that have or are approaching water scarcity should decrease their water consumption. However, the real question is the amount of reduction due to various trade-offs among the three Pillars. Furthermore, in dealing with water (scarcity) the crucial need of nature must be considered particularly under temperature increases, which have a tendency to increase water consumption and consequently scarcity. This is most critical for downstream as a water body approaches its sink. In general, if downstream is healthy then it demonstrates that its water management is credible.

The other two Pillars of water management of a WUS can also promote water scarcity:

- Non-beneficial water consumption, which is an important component of water loss, such as weed evapotranspiration and some of the evaporation from reservoirs and lakes. The idea of non-beneficial is connected to the objectives of a WUS and must be reduced, particularly under water scarcity.
- Non-quality or polluted water, which in effect is a kind of consumption for the WUS under analysis as explained in FIW2a.

FIW2c) Both consumption and inflow should be controlled within the context of integrated trade-offs of the three Pillars.

There are numerous trade-offs between the three Pillars of water management even with valid constraints on each of them. However, water managers look for a solution or combinations of the three Pillars that are reasonable and good-enough for a specified WUS. Clearly, its inflows are crucial because they define the magnitude and nature of water allocation and its influence on water resources. Besides inflows, characteristics of the outflows also need close attention to verify the influences of a WUS on water resources, unrecoverables (FIW2a) and production. In this regard, the consumptive part of the outflow is significant as explained in FIW2a and FIW2b. Therefore, both the inflow and consumptive flows must be carefully analysed and controlled having in mind their comprehensive trade-offs.

Third FIW: Quality

FIW3a) There are water quality norms and regulations for different objectives.

Every objective of a WUS has its own water quality requirements, which should be clearly defined through regulation and legislation. Indeed, it is vital to have proper water quality requirements for public water supply. However, there should be regulations for different categories of water use, such as irrigation, various types of industry and ecosystems. As water scarcity increases, such regulations become essential because of the need to use less freshwater and more of the lower quality water. However, such uses and reuses are subject to social norms, technical expertise, and economic possibilities, meaning that regulations and guidelines are important to start such processes.

FIW3b) Downstream water quality should be protected.

Water pollution and poverty are among the features of the designation upstream/downstream. One of the crucial tasks of water management, or rather its top priority, is to ensure that the water downstream from a WUS has a proper water quality (and quantity). In general, as a river or aquifer approaches a sea or any other sink, pollution increases to unsatisfactory levels, which may happen even if each WUS obeys the legal pollution requirements, but their combined effect do not, particularly under both point and diffuse pollution sources, e.g., see Article 10 of the European Water Framework Directive (European Parliament & Council 2000). At times major water polluters are easy to detect, however enforcing an effective 'polluter pays' mechanism is rather complex.

Dealing with downstream pollution needs financial resources, technical expertise, political will and social acceptance. To advance these needs is hard and can easily fail, however solutions should be envisioned that show progress toward targets in a learning and adaptive mode, which includes the active involvement of all the stakeholders, such as farmers, owners of industries, and urban planners/citizens. It is not acceptable, morally and professionally, that, say after 10 to 15 years, the same downstream problem persists.

FIW3c) Water pollution must be treated adequately for people and nature.

One of the most valuable features of systems is to produce (near) zero water pollution, however this is impractical for almost all the WUSs, hence the critical need for treatment, which requires expertise and investment. Much of the health problems of the people of the world and nature is due to unsatisfactory water and wastewater systems owing to the lack of infrastructures and proper management. First, it is of vital importance to treat the water inflows that are destined for human use. Other water needs of people for activities, such as agriculture, industries, urban green zones, and water for cooling may adequately use water with lower degrees of quality. Additionally, contamination harms the aquatic ecosystems, which in turn may harm human beings, and consequently begs for wastewater treatment and decreasing the use of harmful substances in human activities. Generally, downstream water quality is a good indicator of the effectiveness of controlling water pollution in a WUS, which is an important task of water managers.

Fourth FIW: Benefit

FIW4a) Water benefits should reveal multiple values and costs to all the stakeholders.

Bellagio Principles state "Valuing water means recognizing and considering all the benefits provided by water that encompass economic, social and ecological dimensions" (UN-HLPW 2017). So benefits are not only economical but also reflect social and natural dimensions of life including water equity, conservation and beautification. Benefits are net meaning that the costs of a water project should not be ignored. Costs are crucial to any project and, at least considering all the stakeholders in water, they should be analysed under a wider and deeper understanding than just economics, including the costs of, for example, inequity, no conservation, and ignoring beautification. Sometimes water pricing is used to cover the costs and change behaviour but it is not enough. Maximum water use for households, industries, commerce and irrigation must be defined. Such policies are also very helpful in deciding on possible developments in a region.

FIW4b) The overriding benefit is adequate water supply to all the population.

There is a hierarchy of benefits of adequate (both quantity and quality) water supply, and supplying good water to population has the highest weight. Even though this is obvious to all the decision makers, in practice it is sometimes ignored in favour of various activities that are economically or politically beneficial. In this context, the right of each individual to a minimum amount of water for a dignified living is due to the equality principle accepted and written by all, including the United Nations. Furthermore, water has multiple benefits to different stakeholders, particularly as a human right transmitting the idea that the source of water should be protected for current and future generations.

FIW4c) Investment should benefit multiple water objectives while eliminating allocation extremes.

Any investment in institutions or infrastructures as solutions to water challenges of a WUS should be based on explicit multiple objectives, at least before a final solution is reached. One of the major reasons for the inadequacy of such investments is the lack of comprehensive attention to the extremes of the many descriptors inherent in a WUS and its solutions, such as, satisfactory values of quality parameters and quantity flows, i.e., the idea of equity. It is helpful to remember that almost all the domains of life are in equilibrium under clear minimums and maximums for their various descriptors or variables. In general, there is a smaller range between these two extremes that a system functions the best and as it approaches its extremes, it starts to lose equilibrium and eventually life itself. Even for good things, there must be a maximum. Exaggerating in drinking or eating, even if they are of the best of their kind, makes a person sick. It may not be immediately discernible but if exaggeration continues, in one form or another, balance and health fails. This topic is crucial for the (re)allocation of water quantity as well as its two attributes, and for sustainable efficiency in sustainable equity developed in this book.

Fifth FIW: Demand

FIW5a) Water demand management has priority over water supply infrastructure.

There is a fundamental association between demand and supply, meaning that water management must be carried out from both of these perspectives, in parallel to FIW2c. However, decision making should be centred on a clear and transparent water hierarchy (European Commission 2007, 2018) signifying that additional infrastructures for water supply should be considered after the demand side measures, like water efficiency improvement, real water saving, caps, effective water pricing policy, and cost-effective alternatives. Such a water policy hierarchy is crucial for regions suffering from water scarcity and drought, however it is just good practice and all water managers must implement them in an adaptive environment.

FIW5b) Adaptive water management should advance at multiple levels within a learning approach.

In dealing with the uncertainties of the future, and the complexities of policies and technologies adaptive systems should be planned and engineered. The traditional approaches on averages should be modified in order to engineer systems that are more closely deal with how real situations change. This is because "average of all the possible outcomes associated with uncertain parameters, does not equal (except if system linear) the value obtained from using the average value of the parameters" (de Neufville and Scholtes 2011). Such systems need flexibility, which is "the ability to adapt with minimal cost to a wide range of possible futures. Building in this flexibility may cost more, but may be still desirable insurance against risks society does not want to take" (Loucks and Van Beek 2005). In coming up with such solutions, all the stakeholders should be involved within a learning approach in order to promote equitable and sustainable allocation mechanisms in a transparent manner. Such adaptive systems increase resiliency and reduce risks associated with uncertainty. It should be noted that (a) systems that are based on maladaptive solutions or simply inaction (no solution) may have costs with steeper trade-offs, and (b) adaptation should consider all the possible levels of a systems, i.e., all the positions and extents related to a WUS. The final decisions should be based on bounded rationality as explained here, promoting reasonable and good-enough solutions.

FIW5c) Performance of the interactions between water demand and supply depends on the differentials of its three Pillars.

Description of a WUS is different from its performance, which includes targets. There are two complementary focus into the performance of a WUS; one has to do with combining together the three Pillars, which also reveals integrated trade-offs. The other is about the performance of each WPI and its attributes under specific objectives. These two should be advanced using the following two interrelated types of principles and targets:

- Aggregative: along this policy, the targets should be in percentage (100%) as in the sustainable efficiency (Sefficiency) of a WUS,
- Distributive: according to this policy, the targets are defined in relation to the maximum and minimum values of an indicator of interest for the adequate functioning of the members of a WUS as in the sustainable equity (Sequity).

Quantification of the aggregative performance indicator of a WUS must be based on the differentials of all its three Pillars, i.e., all the inflows and outflows and their beneficial and quality attributes. Such comprehensive and systemic analysis is crucial in order to understand the integrated trade-offs of the three Pillars on a WUS. This means that, it is the difference between what enters into a WUS and what leaves it that gives its performance and not just what enters (e.g., water supply, benefits) or leaves (e.g., water pollution, evapotranspiration). Additionally, the differentials should be employed for both IN and OUT performances, parallel to what is written in FIW2c.

The distributive performance should look at the differential between the most advantaged and the least advantaged members of a WUS in relation to its objectives and targets. Here, the minimum amount of water to safeguard dignity and life, as well as, the maximum values to ensure water and nature conservation should be explicitly defined.

Advancing these principles in the context of the current theory promotes fairness and impartiality. Consequently, the aggregative principle is generally applied after the minimum and maximum requirements are fulfilled.

Before going to the next subsection, let us be clear that this theory sets the minimum foundation upon which Chaps. 4 and 5 are built. Meanwhile, the next two subsections briefly explain three key concerns of the theory, namely, learning, stakeholders and technology.

3.2   Learning with Stakeholders

Ideas matter and evolving good ideas are vital. Any (smart) system starts at a specific level of capacity and grows to higher levels of knowledge, understanding and practice. Think of a six-year-old child going to school to learn which happens year after year with different books and teachers according to her evolution. The same is true with water management of a WUS and stakeholder involvement. "Capacity development is a long-term process… which needs to be continuously adapted according to stakeholders' feedback and needs" (UN-Water 2013). "Transformation of this kind goes beyond performing tasks; instead, it is more a matter of changing mindsets and attitudes" (UNDP 2009). Consequently, two ingredients make the basis of a suitable and continuous learning process: stakeholder's needs, and management's mindsets (Table 3.2). In other words, in advancing water management processes, errors and serious inadequacies occur, however repeating them are signs of lack of comprehensive stakeholder involvement and/or proper mindset of the management, which has to do with attitudes and skills. An acceptable mindset promotes transparency and

Table 3.2 Long-term processes of capacity building of stakeholders and managers

		Mindset	
		Least acceptable	Most acceptable
Needs	Low	Develop a proper learning mechanism along with stakeholders	Progress to new targets with the stakeholders
		A: capacity development of the less advantaged members of the management institution	Capacity and capability development with the three Pillars
	High	Perform A and B given in this table	B: capacity development of the less advantaged stakeholders
		Focus on equity (Chap. 5)	Focus on equity (Chap. 5)

comprehensive stakeholder involvement with learning attitude, starting, say, by the question of Dalio (2017): "Rather than thinking, 'I'm right.' I started to ask myself, 'How do I know I'm right?'"

Table 3.2 (somehow parallel to Fig. 1.3) gives the spectrum of needs and mindsets and, obviously, should not be seen as a binary table. Much is written about capacity development that can be adapted to local circumstances. Here, we broadly try to present some of the fundamental ingredients that should be seriously considered in advancing in a more sustainable water management path.

Table 3.2 shows two types of actors: users (e.g., farmers and city dwellers) and managers (e.g., regional water authorities and boards). The users seek to satisfy their needs with the least cost and are less concern about the influence of their use, meaning that, according to them, the usefulness of their water returns is low. However, the managers have responsibilities beyond a specific user and the returns mostly have high value. This and other differences should be converged through transparent and comprehensive stakeholder participation, having in mind that active involvement itself requires capacity and guidance, particularly for the less advantaged members of the population and institutions as Table 3.2 shows. Furthermore, we should appreciate that any business linked to water has one or two WPI as its focus, while having less interest in the other flows of water balance (Table 2.1). Examples: pump manufacturers emphasize water supply, e.g. VA, and drip irrigation manufacturers talk mostly about ET with some reference to VA. For the former, returns and their qualities are of less importance and for the latter, evaporation (NR) due to improper functioning of the drips is negligible.

It is customary that due to the lack of data, water managers, experts and companies use a flawed indicator or framework, which leads to defective and scientifically inconsistent results. The fact that data does not exist we should not use defective tools, but rather make learning the focus of our water management activities. The core processes in learning are regular *action, reflection and consultation*. Reflection on action is about creating a culture of free exchange of ideas (good and bad) among all the stakeholders (formal or not) on a regular basis. In these crucial meetings, water managers present clearly and with simple language all the experiences of the last period (e.g., four months). Consultation means making decisions by water managers for action based mostly on the ideas and suggestions of the reflection meetings. However, the processes of action, reflection and consultation are about the modes of operation and not a sequence of steps, meaning that many decisions and plans can be taken in the reflection meetings, or even during an action. It should be noted that before any action all the stakeholders should become aware of the decisions and the nature of actions. Today, 'normal' water management reduces the reflection meeting to just one public meeting about a plan instead of being a regular feature of decision-making, which also increases trust (crucial aspect of learning, involvement and management) among the participants. In other words, reflection as the capacity to think, share and analyse collectively should become central in order to increase our understanding about the flow of water among the inhabitants and eventually sustainable development of the region.

"Most people agree that water is an extremely valuable resource" (UN-HLPWb 2017). Almost all the public have great concern or interest (hence stakeholders) about the activities that change the quantity and quality of water, particularly in water scarce and polluted areas (Fig. 3.1). Consequently, "Water management practices will increasingly have to accommodate diversified, even contradictory, demands of stakeholders" (Tortajada and Biswas 2018). They should include experts, decision makers, owners of information, politicians, and various strata of society from poor to rich, disadvantaged to advantaged, industries to farmers, environmentalists to developers, sellers to households, among others (European Commission 2003). Hence, promoting learning within such a diverse setting is indeed a great opportunity for the water managers to bring about positive changes in the development of a region. They can lead sustainable development instead of being one of the many secondary forces of society.

Fig. 3.1 Water stakeholders with different goals and needs (Loucks and Van Beek 2005)

One of the major questions is the role of the water experts and decision makers in relation to other mostly local stakeholders, i.e., the users. In water management, to what extent do the priorities of the experts and the decision makers match that of local stakeholders? The answer to this question is complex, at least in practice. In order to effectively diminish the gap between those priorities the approaches presented in this book require that the nine WPTs in Fig. 2.1 and their two attributes, quality and benefits, be accounted for explicitly in any study, design or plan. To emphasize this requirement, it is important to note that if a particular WPT does not exist or assumed zero, it must be stated explicitly along with the reasons of its exclusion. This leads to a transparent, trusted and stakeholder enabler procedure that strengthens the water management theory of this book. It is obvious that each WPT in Fig. 2.1 has more benefit to a specific number of stakeholders than to others. For example, ET has more importance to farmers, VU to river ecologists, and RP to those interested in returns, such as flows to groundwater. These specific (conflicting) benefits and values are strong enablers for these stakeholders actively participating in managing 'their' water. Hence, the first step to give voice to all the stakeholders is to make the three Pillars of all the WPTs and their WPIs transparent in the format of one summary table. "Making all the values of water explicit gives recognition and a voice to dimensions that are easily overlooked. This is more than a cost-benefit analysis and is necessary to make collective decisions and trade-offs" (UN-HLPW 2017).

Much is written about the crucial need for stakeholder involvement and is part of legal instruments in almost all the countries including the ones in the European Union through the Water Framework Directive (European Parliament & Council 2000). However, rarely the three Pillars, which include all the WPIs and their balance, are revealed explicitly to all the stakeholders under a particular act of management. Besides, most of the experts are interested in a specific area of a Pillar or at maximum in one of them. This partiality and the (presumed) authority of the experts make learning and effective stakeholder involvement more complex. All this is due to "the main constraint being that participation is, in most of the cases, considered as an end by itself instead of as a means to an end" (Tortajada 2010).

Finally, water balance volumes, their qualities and benefits are hard to specify properly for various reasons. Water paths (WPIs) and the degree of their pollution are through measurements, which are generally expensive to perform. Consequently, the first level of capacity development is to produce these data. The idea is to start with a comprehensive estimation of all the WPIs, qualities and benefits of a WUS, and then focus on making those estimates as real as possible in time. The binary mentality that there is data or not should give space to an evolutionary learning environment in data and other water management activities.

3.3   Smart Water Use Systems

Systems, including Water Use Systems, are organised parts or schemes that need two fundamental requirements. First, a scientific and objective theory or framework that presents a comprehensive view of the relationships of all the (major) influencing factors and their trade-offs. Second, reliable data in order to improve the knowledge of those relationships already established. This means that smart data gathering must follow the needs of the theory presented here. Although both of them evolve, in general, the former changes little over time, and most of the latter starts with expert estimates and stakeholder knowledge, both of which become more reliable in time through learning. Besides these two foundational requirements, smart systems should be adaptive and automatic, mostly using technology. Such systems are based on sensing functionality, networking capability, controlling unit, and actuator. Examples of technologies used in smart water systems are sensors to better balance water flows, pollution monitoring and control, remote sensing, smart meters, high tech irrigation systems, digital communication, etc.

"Since we cannot measure what we cannot adequately conceptualize, we have to start with the concept first" (Fukuyama 2013). This is the purpose of this book by presenting a theory that lead to proper measurements for an evolving understanding of the performance of a water use system. In this context, proper technology and smart systems are extremely helpful. However, the priority in using technology should go into finding data that are more accurate for the three Pillars of water management in this order: quantity (WPIs), quality (W_{qX}), benefit (W_{bX}). Of course, in real life most of the things are not binary, so the idea is to go toward comprehensive data accuracy with stakeholders (see Sect. 3.2), technology and water professionals. If this does not happen water management is flawed, having in mind that lack of money should not be the sole excuse as explained in this book.

References

Dalio R (2017) Principles: life & work. Simon & Schuster, sl
de Neufville R, Scholtes S (2011) Flexibility in engineering design. The MIT Press, Cambridge, Massachusetts
European Commission (2003) WFD guidance documents (no. 8). [Online]. Available at: http://ec.eur opa.eu/environment/water/water-framework/facts_figures/guidance_docs_en.htm. Accessed 29 May 2019
European Commission (2007) Addressing the challenge of water scarcity and droughts in the European Union. Communication from the Commission to the European Parliament and the Council, COM/2007/0414 final, Brussels
EuropeanCommission (2018) Regulation of the European Parliament and of the Council on minimum requirements for water reuse. European Commission, Brussels
European Parliament & Council (2000) Water framework directive, Official Journal L 327, European Union. [Online]. Available at: http://ec.europa.eu/environment/water/water-framework/index_ en.html. Accessed 20 Mar 2018

Fukuyama F (2013) What is governance? Center for Global Development, Working Paper 314, Washington, DC

Lankford B (2013) Does Article 6 (factors relevant to equitable and reasonable utilization) in the UN Watercourses Convention misdirect riparian countries? Water Int 38(2):130–145

Loucks D, Van Beek E (2005) Water resources systems planning and management: an introduction to methods, models and applications. UNESCO, Paris

Lu F, Ocampo-Raeder C, Crow B (2014) Equitable water governance: future directions in the understanding and analysis of water inequities in the global South. Water Int 39(2)

Oxford Dictionary (2018) Theory. [Online]. Available at: https://en.oxforddictionaries.com/definition/theory. Accessed 7 April 2018

Rawls J (1999) A theory of justice, Revised edn. Harvard University Press, Cambridge

Tortajada C, Biswas AK (2018) Impacts of megatrends on the global water landscape. Int J Water Resour Dev 34(2):147–149

Tortajada C (2010) Water governance: some critical issues. Int J Water Resour Dev 26(2):297–307

UNDP (2009) Capacity development: a UNDP Primer. United Nations Development Programme. [Online]. Available at: http://www.undp.org/content/dam/aplaws/publication/en/publications/capacity-development/capacity-development-a-undp-primer/CDG_PrimerReport_final_web.pdf. Accessed 22 Mar 2018

UN-HLPW (2017) Bellagio principles on valuing water. United Nations, Sustainable Development Goals (SDG), High Level Panel on Water. [Online]. Available at: https://sustainabledevelopment.un.org/content/documents/15591Bellagio_principles_on_valuing_water_final_version_in_word.pdf. Accessed 29 Sept 2017

UN-HLPWb (2017) Roadmap of the valuing water initiative. Version 1.1. United Nations, Sustainable Development Goals (SDG), High Level Panel on Water. [Online]. Available at: https://sustainabledevelopment.un.org/content/documents/15595Road_Map_Valuing_Water_Initiative_vs_1.1_March_8th_updated_May_19th_2017.pdf. Accessed 29 Sept 2017

UN-Water (2013) Water security & the global water agenda: a UN-water analytical brief. United Nations University, Canada

Chapter 4
Sefficiency (Sustainable Efficiency)

There is a fundamental difference between descriptive and performance indicators of a water use system. The former responds to the question "What is happening?" (e.g., what are the WPIs in a WUS), and the latter focuses on the questions, such as "Does it matter? Are we reaching targets?" (EEA 2015). A target is "a goal to be achieved" (Merriam-Webster Dictionary 2019), and highly important in water management as explained in the previous chapter on the theory presented in this book (FIW5c). For example, ET is part of the description of a WUS (Table 2.1) and by itself value-neutral, because there is no independent target for ET in a locality. As such, decreasing ET in water scarce regions may be a valid option but the real question is the amount of consumption reduction that brings about sustainable development. To answer this and many other questions in water management, we need to use efficiency as a performance indicator with 100 (i.e., in %) as its ideal target. To appreciate efficiency, first let us understand its concept, which is the main reason for its wide usage:

- "Efficiency is thus not a goal in itself. It is not something we want for its own sake, but rather because it helps us attain more of the things we value" (Stone 2012).
- "Resource efficiency means using the Earth's limited resources in a sustainable manner while minimising impacts on the environment. It allows us to create more with less and to deliver greater value with less input" (European Commission 2019).

These two explanations of the concept of efficiency clearly establish its central significance, particularly under water scarcity. It also becomes clear that efficiency should promote sustainability and as will be seen in Chap. 5, it should be an integral part of equity. After understanding the concept, the key question is how to quantify it for a WUS based on the theory presented in this book. Sefficiency developed in

© The Editor(s) (if applicable) and The Author(s), under exclusive license
to Springer Nature Singapore Pte Ltd. 2021
N. Haie, *Transparent Water Management Theory*,
Water Resources Development and Management,
https://doi.org/10.1007/978-981-15-6284-6_4

this chapter fulfils this requirement and if followed properly produces sustainable solutions.

Sefficiency includes economic, environmental and social concerns in its formulation, which makes it a sustainable efficiency indicator for the management of a WUS. However, "sustainability is not a scientific concept, but rather a social goal. It implies an ethic. Public value judgments must be made about which demands and wants should be satisfied today and what changes should be made to ensure a legacy for the future. Different individuals have different points of view, and it is the combined wisdom of everyone's opinions that will shape what society may consider sustainable" (Loucks 2002). This needs transparency and the involvement of all the stakeholders (see Sect. 3.2), because to become sustainable is complex. The quote also refers to "public value judgments", which should be scientifically bounded (Sect. 1.2) as much as possible in order to produce solutions that are more robust for the integrated three Pillars. "In addressing the priority problem the task is that of reducing and not of eliminating entirely the reliance on intuitive judgments" (Rawls 1999). Even though our "priority problem" is different, but the statement also applies to Sefficiency, meaning that it reduces and constraints value judgments significantly.

Before continuing to the proof of Sefficiency, let us present a non-exhaustive list of its possible benefits:

- Enables transparency through the fixed structure of the WUS
- Enables stakeholders involvement for each WPI and its quality and benefit Pillars
- Minimizes the risks of water scarcity
- Adapts to uncertainty (e.g., climate change, population)
- Lessens impact of severe conditions (e.g., drought)
- Protects environment
- Reduces new infrastructure
- Saves energy
- Decreases cost
- Rationalizes investments
- Supports economic growth
- Creates jobs
- Improves cost effectiveness of water service
- Allocates water better
- Enhances conditions for recreation.

4.1 Proof of Sefficiency Indicators

In general, there exists both input and output efficiencies for any system (Coelli et al. 2005) in order to reveal the complexity of its structure and behaviour. Sefficiency equations compute the performance of a WUS from the supply and demand sides (FIW5c). The former is relative to inflow and the latter to consumption, which is a very important part of outflow (FIW2). This proof is based on Haie and Keller

(Macro, Meso, and Micro-Efficiencies in Water Resources Management: A New Framework Using Water Balance 2012) and Haie (Sefficiency (Sustainable efficiency): a Systemic Framework for Advancing Water Security 2013). To start, let us be comprehensive and write the water balance in terms of these two perspectives.

The generic WUS depicted in Fig. 2.1 is composed of Inflow and Outflow with negligible change in storage (FIW1a), which according to the principle of water balance (Inflow = Outflow) can be written as Eq. (4.1):

$$(V1 + OS + PP) - (ET + NR + V2 + RP) = 0 \qquad (4.1)$$

For variable definitions refer to Chap. 2 or the Abbreviations and Symbols in the beginning of this book. Water balance of a WUS can also be presented in terms of consumptive and non-consumptive flows (Table 2.2) as in Eq. (4.2):

$$(V1 + OS + PP - V2 - RP) - (ET + NR) = 0 \qquad (4.2)$$

In order to write an alternative arrangement of water balance that embodies the above two equations and keeps their forms (inflow and consumptive), the binary index ic is introduced, which gives Eq. (4.3):

$$[(V1 + OS + PP) - (1 - ic)(V2 + RP)]$$
$$- [(ET + NR) + ic(V2 + RP)] = 0, \quad ic = \{0, 1\} \qquad (4.3)$$

Note that if $ic = 0$ (i.e., consumptive type), Eq. (4.3) becomes Eq. (4.2), and if $ic = 1$ (i.e., inflow type), we get Eq. (4.1). However, water balance equations are descriptive and do not give any information regarding the performance of a WUS. At this point, we need to introduce the other two Pillars of water management, i.e., quality and benefit attributes into Eq. (4.3). To do so, we apply the Usefulness Criterion given in Sect. 2.2 to Eq. (4.3) as shown in Eq. (4.4):

$$\left\{ \begin{matrix} [(V1 + OS + PP) - (1 - ic)(V2 + RP)]- \\ -[(ET + NR) + ic(V2 + RP)] \end{matrix} \right\}_s = \Lambda, \quad ic = \{0, 1\} \qquad (4.4)$$

Subscript 's' stands for the useful part of all the WPTs and their corresponding WPIs within the curly brackets. This means that $X_S = W_{sX} * X$, with X being a WPI, X_S its useful part, and W_{sX} its Useful Criterion, which is presented in Eq. 2.1. For example, if $X = ET$, then its useful part is $X_S = ET_S = W_{sET} * ET$.

Because X is greater than or equal to X_S and the usefulness of the inflow is more than or equal to the outflow, a non-negative undesirable (Sect. 2.5) factor called Lambda, Λ, is inserted in the right hand side of the equation to maintain the equality. The undesirables of a WPI are non-beneficial and pollution and consequently a fundamental aim of water management is to minimize them. It should be noted that Eq. (4.3) is a special case of the Eq. (4.4), with $W_{bX} = W_{qX} = 1$ (unitary Usefulness Criterion) leading to $\Lambda = 0$, which indicates that there are no undesirables. In a real WUS, these conditions never happen but the idea of good water management is to maximize W_{bX} and W_{qX}. This is to get the highest benefit with lowest pollution possible in a learning environment specific to the WUS under analysis, which in turn minimizes the undesirables. In other words, the basic idea is to [Min (Λ)], which can be written as Eq. (4.5):

$$Min \left\{ \begin{array}{c} [(V1 + OS + PP) - (1 - ic)(V2 + RP)] - \\ -[(ET + NR) + ic(V2 + RP)] \end{array} \right\}_s, \quad ic = \{0, 1\} \qquad (4.5)$$

Minimizing the expression in Eq. (4.5) is equivalent (Appendix A: Equivalency) to maximizing the ratio of its two positive parts as given in Eq. (4.6):

$$Max \left[\frac{ET + NR + ic(V2 + RP)}{V1 + OS + PP - (1 - ic)(V2 + RP)} \right]_S, \quad ic = \{0, 1\} \qquad (4.6)$$

In this book, the expression in Eq. (4.6) that needs to be maximized is called the Sefficiency (SE) equation as presented in Eq. (4.7):

$$SE = \left[\frac{ET + NR + ic(V2 + RP)}{V1 + OS + PP - (1 - ic)(V2 + RP)} \right]_S, \quad ic = \{0, 1\} \qquad (4.7)$$

Using the terminology of Table 2.2, Eq. (4.7) can be written in a more condensed form given in Eq. (4.8):

$$SE = \left[\frac{C + ic * R}{I - (1 - ic) * R} \right]_S = \frac{UC + ic * UR}{UI - (1 - ic) * UR}, \quad ic = \{0, 1\} \qquad (4.8)$$

Sefficiency definitions according to the two perspectives and in relation to the last two equations are:

- iSE = Inflow Sefficiency ($ic = 1$): the ratio of useful Outflow to useful Inflow
- cSE = Consumptive Sefficiency ($ic = 0$): the ratio of useful Consumption to Total Unrecoverable Flow (TUF$_S$)

Generally, SE is multiplied by 100 to give a percentage, making it a positive number that has to be less than 100, meaning that the denominators must be greater than the corresponding numerators (if not, data may be inaccurate or unrealistic). It should be noted that Sefficiency rarely, if ever, is 100%, because it is very costly, even impossible, to have a WUS without undesirables (see Sect. 2.5).

There is the possibility of calculating Sefficiency without any quality consideration, because it is informative to see the difference between the Sefficiency (SE) of a WUS and its beneficial Sefficiency (SE$_b$), i.e., one without water quality (W_{qX} = 1). Although, we suggest that SE and not SE$_b$ be the main indicators for water management, Eq. (4.9) gives SE$_b$ by using Eq. (4.7):

$$SE_b = \left[\frac{ET + NR + ic(V2 + RP)}{V1 + OS + PP - (1 - ic)(V2 + RP)} \right]_b, \quad ic = \{0, 1\} \qquad (4.9)$$

$$SE_b = \left[\frac{C + ic * R}{I - (1 - ic) * R} \right]_b = \frac{BC + ic * BR}{BI - (1 - ic) * BR}, \quad ic = \{0, 1\} \qquad (4.10)$$

A caution needs to be exercised as explained in Sect. 2.2 in applying weights to the sum or difference of the flows. For example, if $ic = 0$, the inside of the square brackets of Eq. (4.8) and Eq. (4.10) becomes C/ (I − R). However, Eq. 2.2 shows that I − R = C, hence making the inside of the brackets always equal to C/C = 1 and SE = $[1]_s$ = 100%. Of course, this is not correct because, as was explained in the said subsection, first we have to apply the weights to each element inside the brackets and then make the arithmetic operations.

4.2 Levels of Management

Multilevel water management (FIW5) is central in understanding the various impacts and trade-offs of a WUS. Section 2.2 presented three levels for a WUS, viz.: Macro, Meso and Micro, which are defined according to the flows considered. Sefficiency is applied to these three levels and are called 3ME (= Macro-, Meso-, Micro-efficiencies) with their schematics shown in Fig. 4.1.

1. Macro Sefficiency (MacroSE)

   a. The main source of water, and consequently the basin is considered.
   b. Condition of the main source of water, e.g., a river, influences Sefficiency.
   c. MacroSE reveals the effect of WUS on the main source of water.
   d. For Eq. (4.7) to Eq. (4.10) V1 = VU and V2 = VD.

2. Meso Sefficiency (MesoSE)

   a. MesoSE ignores the main source of water.
   b. The prefix meso means "Middle; intermediate" (Oxford Dictionary 2018). Usually, meso comes with micro and macro meaning something between these two.
   c. MesoSE reflects the effect of a WUS on downstream by considering its returns.
   d. For Eq. (4.7) to Eq. (4.10) V1 = VA and V2 = RF.

3. Micro (MicroSE)

   a. MicroSE does not consider the main source of water nor the returns of the WUS. This means that MicroSE ignores the effects on downstream with iMicro = cMicro.
   b. Micro is about the flows or their proxies (e.g., Euros) of direct interest to the stakeholder (e.g., farmer, factory owner, ecosystem NGO, city planner).

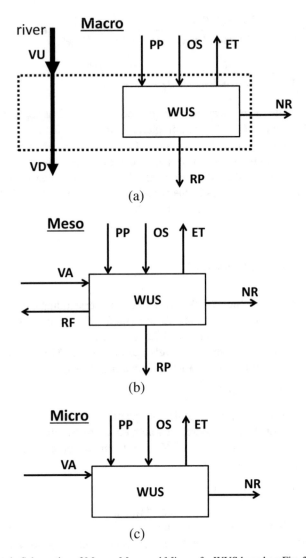

Fig. 4.1 Generic Schematics of Macro, Meso and Micro of a WUS based on Fig. 2.1

   c. It is not based on water balance and as a result prone to errors from the point of view of water management.

   d. For Eq. (4.7) to Eq. (4.10) $V1 = VA$ and $V2 + RP = 0$.

     The dotted rectangle in Fig. 4.1 is for better visualization and indicates a construct showing the flows that are part of the Macro level analysis. Substituting V1 and V2 into Eq. (4.7), we get the following Sefficiency equations:

$$MacroSE = \left[\frac{ET + NR + ic(VD + RP)}{VU + OS + PP - (1 - ic)(VD + RP)} \right]_S, ic = \{0, 1\}$$

$$MesoSE = \left[\frac{ET + NR + ic(RF + RP)}{VA + OS + PP - (1 - ic)(RF + RP)} \right]_S$$

$$MicroSE = \left[\frac{ET + NR}{VA + OS + PP} \right]_S \tag{4.11}$$

Equation (4.11) gives the following five Sefficiency definitions:

- iMacroSE = Inflow MacroSE$_S$ ($ic = 1$): the ratio of useful Macro-Outflow to useful Macro-Inflow
- cMacroSE = Consumptive MacroSE$_S$ ($ic = 0$): the ratio of useful Consumption to Macro-TUF$_s$
- iMesoSE = Inflow MesoSE$_S$ ($ic = 1$): the ratio of useful Meso-Outflow to useful Meso-Inflow
- cMesoSE = Consumptive MesoSE$_S$ ($ic = 0$): the ratio of useful Consumption to Meso-TUF$_s$
- iMicroSE = cMicroSE = MicroSE: the ratio of useful Consumption to useful Micro-Inflow

Examples for the above terminologies:

- Macro-Outflow = ET + NR + VD + RP
- Meso-TUF = VA + OS + PP – (RF + RP)
- Micro-Inflow = Meso-Inflow = VA + OS + PP

Based on Eq. (4.9), the following beneficial 3ME equations can be written:

$$MacroSE_b = \left[\frac{ET + NR + ic(VD + RP)}{VU + OS + PP - (1 - ic)(VD + RP)} \right]_b, ic = \{0, 1\}$$

$$MesoSE_b = \left[\frac{ET + NR + ic(RF + RP)}{VA + OS + PP - (1 - ic)(RF + RP)} \right]_b$$

$$MicroSE_b = \left[\frac{ET + NR}{VA + OS + PP} \right]_b \tag{4.12}$$

Equation (4.12) gives the following five bSefficiency definitions:

- iMacroSE$_b$ ($ic = 1$): the ratio of beneficial Macro-Outflow to beneficial Macro-Inflow
- cMacroSE$_b$ ($ic = 0$): the ratio of beneficial Consumption to Macro-TUF$_b$
- iMesoSE$_b$ ($ic = 1$): the ratio of beneficial Meso-Outflow to beneficial Meso-Inflow
- cMesoSE$_b$ ($ic = 0$): the ratio of beneficial Consumption to Meso- TUF$_b$
- MicroSE$_b$: the ratio of beneficial Consumption to beneficial Micro-Inflow

Equation (4.11) and Eq. (4.12) give eight important indicators, namely, iMacroSE, cMacroSE, iMesoSE, cMesoSE, iMacroSE$_b$, cMacroSE$_b$, iMesoSE$_b$, cMesoSE$_b$.

They have 56 distinct combinations, but only 12 of them in three impact categories are important and will be covered in the following Sect. 4.4.2 (Three Impacts in Differentials of Sefficiency). The two Micro efficiencies are flawed as mentioned above in this subsection, with MicroSE$_b$ being analogous to the flawed Classical Efficiency explained in the Sect. 4.5 below.

Real cases are generally one to many and consequently have many WPIs. Here, two templates are available: (a) a free MS-Excel tool at Haie (Sefficiency (Sustainable efficiency) of Water-Energy-Food Entangled Systems 2016) as a supplementary document that has the equations and a simple format to simulate various cases side by side, (b) Appendix B: Sefficiency Template is the compact form used for this book. Although these are one to one templates (i.e., WPT $\rightarrow$ WPI), they can be edited to accommodate one to many scenarios.

Finally, in many applications, initially, the water managers can use Eq. (4.8) and Eq. (4.10). This reduces the needed water quantities to two, most of the time I and C, and utilizing the water balance equation, i.e., Equation 2.2, R can be found. These three water quantities need six weights for their quality and beneficial attributes to be able to calculate Sefficiency.

4.3 Weights

Most of the fields of science, if not all, employ weights explicitly and implicitly. They may be difficult or controversial to set but as the Nobel Laureate Amartya Sen (Inequality Reexamined 1992) affirms "weighting cannot really be, in any sense, an embarrassment" and "We cannot criticize the commodity-centred evaluation on the ground that different commodities are weighted differently." Prices are weights and we readily accept that different commodities have different prices in the same location, and the same commodity has different prices in various locations.

In this book, there are two weights for each WPI (W_{bX} and W_{qX}), which are set according to the objectives of the WUS under analysis. They are explained in the next two subsections, along with a subsection on Usefulness Criterion.

4.3.1 Quality Attribute

"Most economic systems refer to pollution as an "externality;" a cost or benefit unaccounted for in the economic system. Pollution is a negative externality. Anyone taking rudimentary economics should know this. Solutions that have been working in the U.S. and many other places in the world involve government regulation of polluters. Take these externalized costs and integrate them into the system so that humans and aspects of nature that are suffering from their negative impacts without receiving the benefits are protected" (Fitch 2012).

To do this, Sefficiency uses quality weight for each WPI (W_{qX}) in order to explicitly quantify pollution influence on the performance of a WUS. Sometimes it is easy to set those weights:

- Quality weight for treated water supplied to population is one ($W_{qI} = 1$).
- Quality weights of evapotranspiration (W_{qET}) and some of non-reusable (W_{qNR}), such as evaporation and bottled water are one.

However, in most situations, quality weight of a WPI is less than one and should be calculated using estimates and measurements having in mind that a water quality index (or quality weight) "is a weighted average of selected ambient concentrations of pollutants usually linked to water quality classes" (OECD 2001). What follows is a small list that can guide in setting quality weights:

- The Water Framework Directive (European Parliament & Council 2000) introduced surface water status and groundwater status, which are general expressions of the status of a body of surface water or groundwater, respectively. Both include chemical status and classify waters into clearly defined categories with associated colours.
- Canadian water quality guidelines for the protection of aquatic life employs Water Quality Index (CCME 2017), which has a calculator and a user's manual.
- The main global water quality index for domestic purposes is the Global Drinking Water Quality Index of the Global Environmental Monitoring System (GEMS) Water Programme, the United Nations Environment Program (UNEP 2007), and is based on the Canadian Water Quality Index.
- Chinese Environmental Quality Standards with five classes are "formulated for implementing the Environmental Protection Law and Law of Water Pollution Prevention and Control of People's Republic of China, and to control water pollution and to protect water resources" (MEEC 1997; MEEC 2018).
- To calculate water footprint, Hoekstra et al. (The water footprint assessment manual: Setting the global standard 2011) introduced grey water footprint as "an indicator of freshwater pollution" in order to achieve a water quality objective/standard. In this context, grey water footprint may be employed to set W_{qX} under some conditions.
- Food and Agriculture Organisation of the United Nations proposes the use of leaching fraction (Ayers and Westcot 1994) that is widely employed in irrigation management to avoid salt accumulation and can be used for setting quality weight ($W_{qX} = 1 - LF_X$, LF being leaching fraction).

4.3.2 Beneficial Attribute

The beneficial weight (W_{bX}) is set by focussing on the nature and objectives of the WUS under consideration without considering its quality ($W_{qX} = 1$). This is the usual way that the planning and management of the systems are carried out today, i.e., water quality of a system is dealt with separately according to its inflow needs and the downstream requirements. This weight should consider all the benefits that water brings to societies and natures: "Valuing water means recognizing and considering all the benefits provided by water that encompass economic, social and ecological dimensions. It takes many forms appropriate to local circumstances and cultures. Safeguarding the poor, the vulnerable and the environment is required in all instances" (UN-HLPW 2017). Hence, water benefits (interchangeable with water values according to this citation and others) are fundamentally linked to the objectives of a WUS, and include those that may not be quantifiable.

FIW4 is about the benefits, which stresses that multi-objective planning and management should be the norm. Total benefits and costs vary due to various local or national goals and even under different water allocation schemes. There are many methods available to come up with those totals (Loucks and Van Beek 2005) and then the weights. Presentation of these methods are beyond the scope of this book, but they are routinely utilized in evaluating economic and social benefits of projects. However, in many cases a competent estimate of the magnitude of the weight is sufficient, which may be valuable as the first step in a learning process. At present, the following is common practice in applications all over the world:

- Public water supply to people has a W_{bI} of one, meaning that all the water that enters into the water supply system (Inflow to a WUS) has the maximum benefit.
- For irrigation systems, the so-called effective precipitation (Brouwer and Heibloem 1986) can be used to set the beneficial weight of precipitation (W_{bPP}). This is doubly important for rainfed agriculture.
- The non-beneficial ET is routinely estimated or calculated at least for irrigation systems.
- Evaporation from lakes and reservoirs are calculated with a small fraction considered as beneficial.

Finally, gathering accurate data for the management of water systems is very hard due to many factors, such as, bias (different from prejudice) and noise (chance variability of judgments), shown in Fig. 4.2 (Kahneman et al. 2016). Noise is one of the reasons that a learning process is needed because sometimes what a stakeholder presents under one situation may be different from what he expresses under another (consciously or not). Furthermore, decision makers and politicians act with much noise for advancing their interests, which make the stakeholders even noisier. However, action (Sect. 3.2) truly reveals the real intentions and true mindsets (Sect. 3.2) of all involved in water management (or life in general). Due to such inherent conditions and the idea of bounded rationality (Sect. 4.2), it is suggested that equal weights (or its special case, unit weights) can sometimes be justifiable for complex systems such as, the NASDAQ-100 Equal Weighted Index Shares (NASDAQ-100). This may be used for the beneficial weights (W_{bX}) at least in situations that reliable data is not available, an initial estimate is needed in the learning process, or under urgent situations. Although the problem of noise is explained here for W_{bX}, but it is applicable for all types of data, including quality and quantity, and should be persistently dealt with in all data handling.

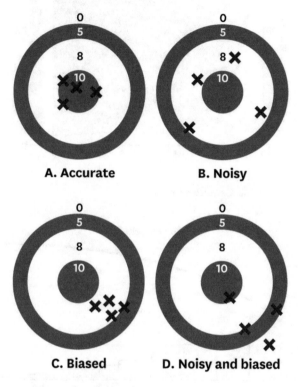

Fig. 4.2 Noise and bias in data accuracy (Kahneman et al. 2016)

4.3.3 Usefulness Criterion

The quality and beneficial weights vary between 0 and 1 as discussed earlier, and their multiplication defines Usefulness Criterion (Sect. 2.2), which is an equation in the form of $z = x * y$. Figure 4.3 gives the contour lines of the domain of W_{sX} generated by MatLab (MathWorks 2018). This figure shows the non-linear behaviour of W_{sX} and that its values are mostly low. In fact the average of all the points that formed the figure is just 0.25.

Figure 4.4 shows the histogram and cumulative curve of W_{sX} using the data produced for Fig. 4.3. The former presents the percentages of W_{sX} values in each bin or class (e.g. 33.5% in class 0 to 0.10), and the latter shows the percentages that are

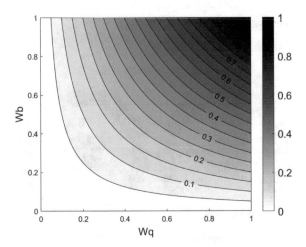

Fig. 4.3 Contour lines of the Usefulness Criterion

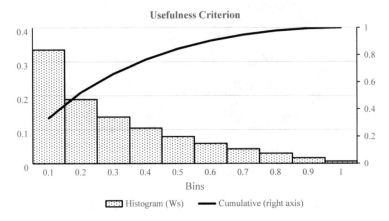

Fig. 4.4 Histogram and cumulative curve of the Usefulness Criterion

less than or equal to a particular value (e.g. 66.1% for 0.3). It is interesting to note that 84.5% of W_{sx} values are less than or equal to 0.5, 90.5% to 0.6, and 97.7% to 0.8, which means achieving high usefulness is difficult.

There is yet another important behaviour of the Usefulness Criterion by following the Benford's Law (Weisstein 2018; Berger and Hill 2015). It gives a fixed probability distribution of the leading nine significant digits, i.e., one to nine, of many types of collections of numbers, such as, river areas and population. The Benford's Law (or the Newcomb–Benford law) has a logarithmic distribution as given by Eq. (4.13).

$$P(g) = log_{10}\left(1 + \frac{1}{g}\right), \; for \; all \; g = 1, 2, 3, \dots, 9 \tag{4.13}$$

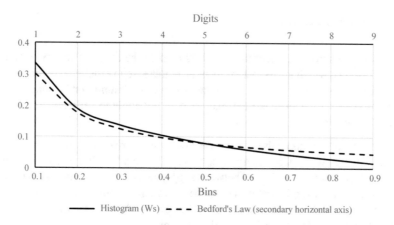

Fig. 4.5 Curves of Usefulness Criterion histogram and Benford's Law

P(g) is the probability of g, which is the leading significant digit (non-zero). Figure 4.5 shows the two curves: histogram of the Usefulness Criterion and Benford's Law. Their differences are negligible, which go from 0.03 to -0.03 with a zero average.

These patterns seem significant particularly parsing the quality and beneficial weights in relation to Sefficiency and in the context of Benford's Law. For example, the streamflow data so important to the water balance and hence the theory presented in this book should conform to the Benford's Law and nonconformity could indicate specific issues with the data (Nigrini and Miller 2007). In other words, if the Usefulness Criterion in a locality diverges from the Benford's law, it should raise a flag as to its accuracy, meaning that the beneficial and quality weights should be re-examined. However, the methodology to actually finding those divergences among many beneficial and quality weights is not clear. Nevertheless and beyond what was mentioned earlier in Sect. 2.2, these findings are indicative that the Useful Criterion defined in this book is sound and valid.

4.4 Trade-Offs

Trade-offs between the three Pillars of water management are inevitable particularly under water scarcity and are highly complex to quantify. Sefficiency as a centrepiece of the theory advanced in this book is about that complexity, meaning achieving a better trade-off and consequently reducing the undesirables. However, water resources development of an area has limits, i.e., the trade-offs of the three Pillars of water management have a reasonable upper bound (Sect. 1.2) in the context of the performance of systems. This feature can be used to reject the development plans, such as a new industrial plant, park or irrigated farm that reduce the performance of the system.

4.4.1   Jevons Paradox

Some water experts in presenting trade-offs refer to the Jevons Paradox (a type of rebound effect), which states that "if there is an increase in efficiency in the use of a resource its price can reduce, leading to an increase in consumption" (Maxwell et al. 2011). However, such an economic analysis does not apply to Sefficiency in light of the following reasons:

- Jevons Paradox is about those resources that have one state after usage and not two like water (FIW2a). Coal energy (the focus of Jevons Paradox of 1865) presents possible paradoxical trade-offs between supply side efficiency, demand (one state) and price. However, in water management, we have iSefficiency, water demand (two states), and price that reveal more complexity than energy efficiency, some of which are made clearer in the following points.
- Various local and global drivers are increasing water scarcity meaning that effective supply is decreasing, which does not allow the price of water to decrease (actually the prices are increasing almost everywhere). These are not the underlying assumptions for Jevons Paradox.
- Jevons Paradox does not consider pollution but Sefficiency does.
- The solutions according to Sefficiency do not necessarily increase water demand or reduce its price because of the complex trade-offs of the three Pillars.
- The use of technology in production processes of energy was another focus of Jevons Paradox. However, technology in Sefficiency is for data gathering in a learning process in order to better estimate the three Pillars, and make water balance (FIW1) more robust. In general, this increases the cost of water supply (not decreasing according to Jevons Paradox) but eventually makes planning and management of this vital resource more sustainable.
- In the absence of proper policies, it is possible that production technologies, e.g., in irrigation, cause water consumption to increase. However, this does not mean that Sefficiency increases because of the trade-offs of the three pillars. In other words, it is not water consumption alone but rather the performance of the WUS within a specific situation that is the deciding factor. This is again different from the logical setting of Jevons Paradox.

4.4.2   Three Impacts in Differentials

The eight important Sefficiency indicators (Sect. 1.2) give twelve significant combinations that show trade-offs between those indicators. These are divided into three impact categories as follows:

- I/O impacts are due to the differences between inflow and consumptive Sefficiencies at the same Level and Pollution. This is done via the following four comparisons:

- • iMacroSE and cMacroSE
- • iMesoSE and cMesoSE
- • iMacroSE$_b$ and cMacroSE$_b$
- • iMesoSE$_b$ and cMesoSE$_b$

- • Level impacts are due to the differences between Macro and Meso Sefficiencies at the same I/O and Pollution. This is done via the following four comparisons:

  - • iMacroSE and iMesoSE
  - • iMacroSE$_b$ and iMesoSE$_b$
  - • cMacroSE and cMesoSE
  - • cMacroSE$_b$ and cMesoSE$_b$

- • Pollution impacts are due to the differences between full and beneficial Sefficiencies at the same I/O and Level. This is done via the following four comparisons:

  - • iMacroSE and iMacroSE$_b$
  - • iMesoSE and iMesoSE$_b$
  - • cMacroSE and cMacroSE$_b$
  - • cMesoSE and cMesoSE$_b$

Under a specific application, we should start with the worst impact difference and analyse it in more detail in conjunction with the trade-off patterns given in the next subsection. In general, these repeating reflections with stakeholders can progress to unconventional scenarios that sometimes can disrupt the usual functioning of a WUS. In other words, having water as the main priority, not the traditional ones, such as, economy, food, land, health or ecosystem, can lead us to innovative scenarios for the sustainable development of a region.

4.4.3 Patterns

Equation (4.7) has more than 13 variables, which are WPIs and their weights. Changing one variable gives a different value for SE in a mostly non-linear fashion. Frequently, if one variable changes, others also vary making the combined effect of all the changes on SE more difficult to assess, meaning that trade-offs are much more complex. In general, the notion of trade-off in this subsection is to understand the behaviour and the structure of the domain (or space) of a policy or in this situation Eq. (4.7). Please see the end of Sect. 2.1 for the clarification of the notion of domain behaviour of an equation.

However, Eq. (4.7) is very complex because of its high dimensions, so let us see the behaviour of Eq. (4.8), which has six variables (I, C, R and their Usefulness Criteria). To start and remembering Eq. 2.2, we define the expressions given in Eq. (4.14).

$$C1 = \frac{C}{I}, R1 = \frac{R}{I}, C1 + R1 = 1$$

$$WC1 = \frac{W_{dC}}{W_{dI}}, WR1 = \frac{W_{dR}}{W_{dI}}, d = b, q, s \tag{4.14}$$

C1 is Consumption fraction, R1 is Return fraction, WC1 is desirable Consumption fraction, and WR1 is desirable Return fraction. For example, if d = b, WC1 is beneficial Consumption fraction. Applying Eq. (4.14) to Eq. (4.8), we get Eq. (4.15).

$$SE = \frac{WC1 * C1 + ic * WR1 * R1}{1 - (1 - ic) * WR1 * R1}, ic = \{0, 1\} \tag{4.15}$$

Equation (4.15) and C1 + R1 = 1 form a 4D problem: C1 or R1, WC1, WR1 and SE (for d = b, we should use the symbol SE_b). This is still difficult to visualize, but keeping one of the variables constant, contour lines of iSE and cSE can be shown as in Figs. 4.6, 4.7, 4.8, 4.9, 4.10, 4.11, 4.12, 4.13, 4.14, 4.15, 4.16 and 4.17 (MatLab (MathWorks 2018) was used, with special thanks to Rui M.S. Pereira). These figures give trade-offs and patterns between the variables.

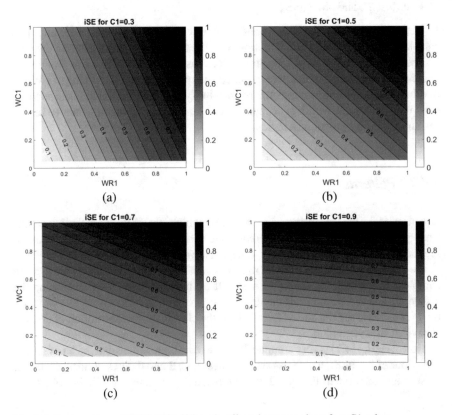

Fig. 4.6 Contour lines of WR1-WC1-iSE trade-offs and patterns along four C1 values

Note: each figure shows four graphs in the x-y-z format with the following characteristics:

- Horizontal axis shows C1, R1 or WR1
- Vertical axis shows WC1 or WR1
- Within these axes, contour lines of IN Sefficiency (iSE) or OUT Sefficiency (cSE) are drawn
- Each figure shows various graphs along C1, R1, WC1 or WR1. We give four out of 20 graphs, which seems to be sufficient to portray the patterns and the trade-offs.

For example, Fig. 4.6 gives WR1-WC1-iSE graphs with WR1 and WC1 axes showing iSE contour lines for four fixed values of C1.

It should be mentioned that not all the combinations of the three variables (C1 or R1, WC1, WR1) occur in practice, meaning that there are infeasible combinations within each figure. However, the patterns and trade-offs given by the figures are valid for the feasible combinations, and consequently, in the following discussions, we will not be concerned about the feasibility of an example or case (i.e., one specific

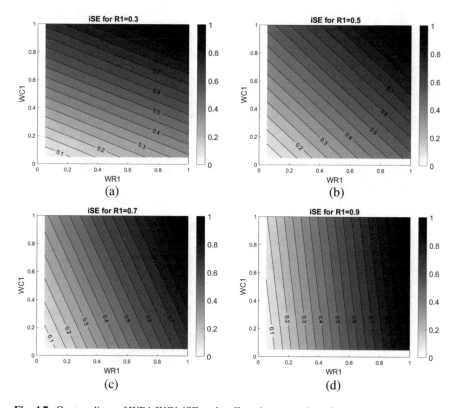

Fig. 4.7 Contour lines of WR1-WC1-iSE trade-offs and patterns along four R1 values

combination of the three variables), but rather how to detect patterns and trade-offs. Furthermore, the figures show the domain (space or hyperspace) of Eq. (4.15), meaning that whatever is mentioned below can be proven by this equation having in mind that $C1 + R1 = 1$. In this context, let us see a non-exhaustive list of the patterns and trade-offs inherent in the figures, which can help water managers discussing their own conclusions for their cases.

- Because $C1 + R1 = 1$, specifying $C1$ or $R1$ will automatically fix the other, which produces equal graphs. For example, the graph for $C1 = 0.3$ in Fig. 4.6 is the same as the one for $R1 = 0.7$ of Fig. 4.7.
- For Figs. 4.6, 4.7, 4.8 and 4.9, the contour lines are linear and for any graph the Sefficiency increases as $WR1$ and/or $WC1$ increases.
- SE sometimes shows high variation or gradient. For example, in Fig. 4.9c, as cSE increases, the contours get closer to each other.
- Various figures mirror each other. For example, Figs. 4.11 and 4.13 are mirrors of Figs. 4.10 and 4.12, respectively, giving equal SE for the complimentary $C1$ and $R1$ graphs. For example, $iSE = 64\%$, for $C1 = 0.8$, $WR1 = 0.4$ in Fig. 4.10c, and

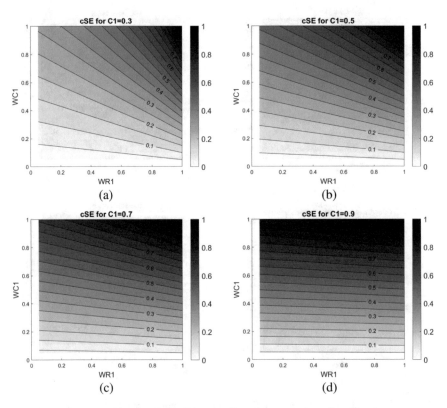

Fig. 4.8 Contour lines of WR1-WC1-cSE trade-offs and patterns along four C1 values

R1 = 0.2 (= 1-C1), WR1 = 0.4 in Fig. 4.11c. For cSE = 60% and WR1 = 0.5, C1 = 0.75 in Fig. 4.12c, and R1 = 0.25 in Fig. 4.13c.

- iSE decreases toward its minimum (zero) if WC1 and WR1 decrease toward zero (on the limit, iSE = WC1 = WR1 = 0). WC1 = 0 means totally useless Consumption or C1 = 0. WR1 = 0 means totally polluted Return or R1 = 0.
- iSE goes toward WC1 if C1 goes toward one, or WC1 and WR1 get closer to each other (on the limit, iSE = WC1 = WR1).

    - If in a real case WC1 is always greater than WR1, which seems to be a valid condition in most of the situations if not all, then the maximum that iSE can achieve is WC1.
    - If WC1 < WR1, then iSE > WC1; If WC1 > WR1, then iSE > WR1.

- iSE goes toward WR1 if R1 goes toward one.
- iSE goes toward C1 if WC1 goes toward one and WR1 goes toward zero.
- cSE decreases toward its minimum (zero) as WC1 and/or C1 goes toward zero.
- cSE increases toward its maximum (WC1) as WR1 and/or C1 goes toward one.

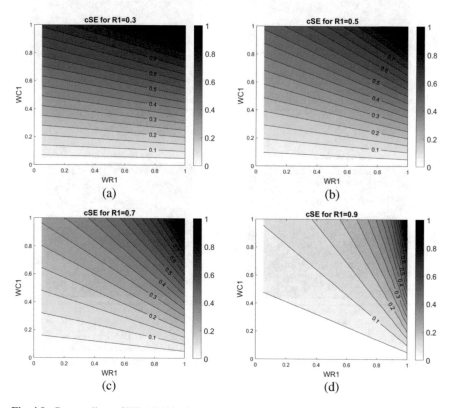

Fig. 4.9 Contour lines of WR1-WC1-cSE trade-offs and patterns along four R1 values

- cSE and iSE go toward each other, if WR1 or R1 goes toward zero. In general, iSE and cSE go toward each other as they get closer to WC1 * C1, which at the limit, we have iSE = cSE = WC1 * C1.

4.5 Alternatives

There are alternative indicators that are also about computing efficiency, productivity, etc. One significant form is to specify the ratio of output to input. Output gives the things that we value, which goes along our objectives, and input is a sort of total. Here, we briefly discuss four alternatives, namely, Classical Efficiency (CE), Water Productivity (WaP), Effective Efficiency (EE), and also Resiliency (RE).

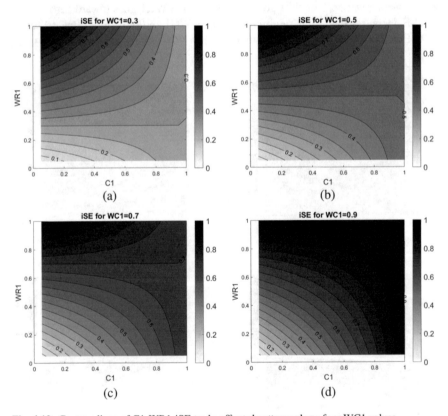

Fig. 4.10 Contour lines of C1-WR1-iSE trade-offs and patterns along four WC1 values

4.5.1 Classical Efficiency

All sectors, including urban and agriculture, have been employing Classical Efficiency (CE) for decades with Eq. (4.16) giving its generic form.

$$CE = \frac{Beneficial\ water}{Total\ water} \tag{4.16}$$

This is to say that CE is the ratio of beneficial water to total water (applied). Examples of the word 'water' in Eq. (4.16) are as follows:

- Numerator (beneficial water): beneficial water use, consumption or required
- Denominator (total water): total water applied, abstracted, allocated or required

With such a definition, CE is a flawed indicator, having in mind the following issues:

- Incompleteness of water flows, particularly the lack of the inclusion of returns in the equation, which also makes it inadequate for any multi-objective WUS.

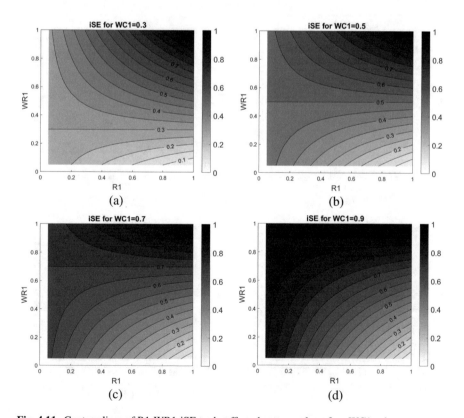

Fig. 4.11 Contour lines of R1-WR1-iSE trade-offs and patterns along four WC1 values

- CE does not obey water balance, one of the most important laws in studying and designing water systems.
- Partial consideration of Usefulness Criterion, i.e., lack of a comprehensive concern for applying water quality and benefits. For example, the numerator considers the beneficial part, but this distinction is not extended to the denominator. Furthermore, CE is a quantity indicator, with little or no consideration for water quality.

Many experts state that increasing CE by decreasing the total water (denominator) saves water for downstream. However, in most cases such a water saving is actually negligible and close to zero. Please see a common example about this myth in Chap. 6. For now, let us focus on the application of CE in agriculture and urban areas.

In irrigated agriculture, CE is mostly defined as $CE = ET_b/VA$ (Seckler et al. 2003), meaning the ratio of beneficial ET to water applied. Various other names are given to CE, such as, irrigation efficiency and water use efficiency. As just mentioned CE is flawed and many authors, including Willardson, et al. (1994) and Haie and Keller (2012) have discussed its problems. Additionally, various authors have defined CE with some variations but all are flawed. For example, efficiency is shown as 1/CE

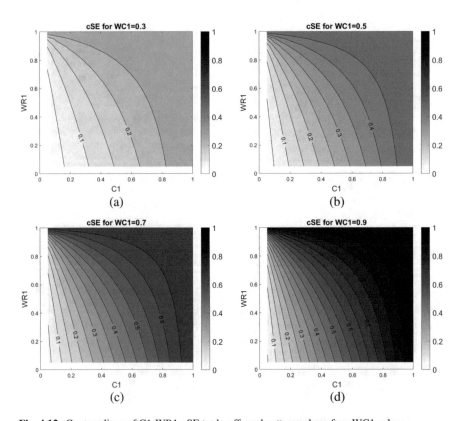

Fig. 4.12 Contour lines of C1-WR1-cSE trade-offs and patterns along four WC1 values

and given a different name, or leaching fraction is applied to VA (see Sect. 1.3), or PP_b (also called effective precipitation) is subtracted from ET_b, or change in storage is subtracted from VA (in practice never used, e.g., Burt et al. (1997)), etc. In reality, CE is a fraction, which conveys very little information (Willardson et al. 1994) and not a scientific and logical performance indicator. However, CE is, to some extent, legitimate from the perspective of the crop but not water, meaning that for agronomists, CE may advance some information but for water managers, it is not suitable at all. These two perspectives create confusion in the mind of many experts, making it another example of why this book insists on the centrality of water in managing it.

In urban areas, the concept of CE, i.e., Eq. (4.16), is also wide spread and equally flawed due to the three issues mentioned above. Let us see three examples by focusing on concepts (index 1 = before intervention, and index 2 = after intervention for CE improvement):

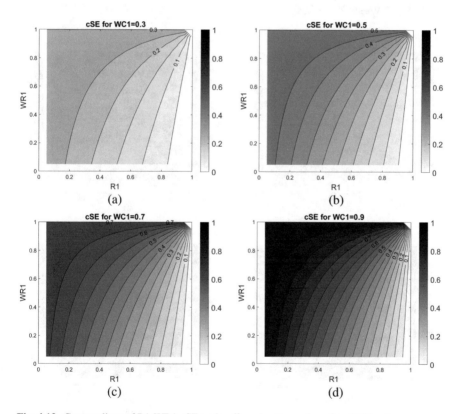

Fig. 4.13 Contour lines of R1-WR1-cSE trade-offs and patterns along four WC1 values

First, the California Department of Water Resources (CDWR) defines urban water use efficiency as "Methods or technologies resulting in the same beneficial residential, commercial, industrial, and institutional uses with less water or increased beneficial uses from existing water quantities" (CDWR 2019). Carefully reading this definition, we reach the conclusion that it is about an expression with its numerator being beneficial water use (X_b), and its denominator, total water quantity applied (VA), which represents Eq. (4.16). To understand this affirmation, let us enumerate the possibilities that the definition sets forth to improve efficiency:

A. $X_{b1} = X_{b2}$ with $VA_2 < VA_1$ (this is the 1st part of the CDWR definition of urban efficiency)
B. $X_{b1} < X_{b2}$ with $VA_2 = VA_1$ (this is the 2nd part of the definition, which is after "or")
C. A third possibility is not presented in the above CDWR definition of urban efficiency with $X_{b1} \neq X_{b2}$ and $VA_2 \neq VA_1$. Under these conditions, CE improves if $X_{b1} * VA_2 < X_{b2} * VA_1$.

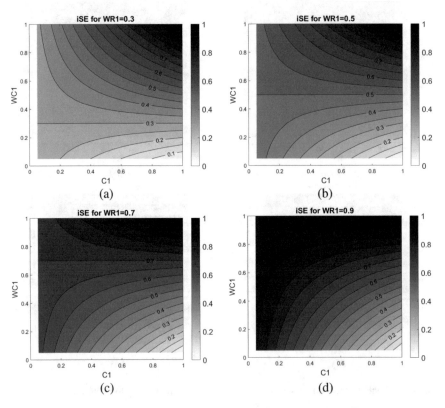

Fig. 4.14 Contour lines of C1-WC1-iSE trade-offs and patterns along four WR1 values

Second, the European Union has defined a number of building blocks for water efficiency activities (European Commission Water 2019) which highlights the following two points, among others:

- "A study from 2007 on the water saving potential in Europe, estimates that water efficiency could be improved by nearly 40%". This study (Dworak 2007) was funded by the European Commission, Directorate-General Environment and produced a report that goes into many sectors and services. Its overall outlook defines water saving potential stating that it "can be achieved by improving the efficiency of various uses of water without decreasing services or by cutting back the use of a resource, even if that means cutting back the goods and services produced by using that resource." This definition gives two possibilities, which Eq. (4.16) should be used for their understanding: (i) $CE_2 > CE_1$ and $X_{b2} \geq X_{b1}$, (ii) after "or", $VA_2 < VA_1$ and $X_{2b} \leq X_{b1}$ with an implicit assumption that $CE_2 \geq CE_1$.
- "The study identifies the need for an EU approach that could contribute to water efficiency across Europe, regardless of the variation in climate, population or land use practices in Member States." This study on water efficiency standards

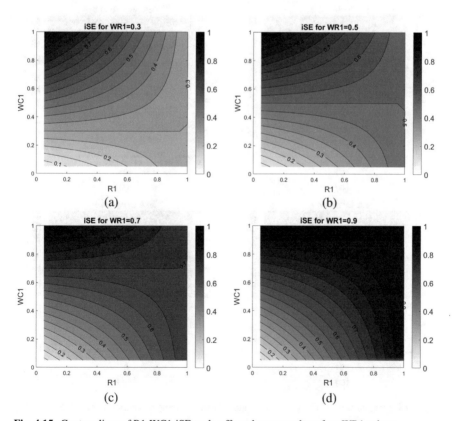

Fig. 4.15 Contour lines of R1-WC1-iSE trade-offs and patterns along four WR1 values

(Benito 2009) was funded by the European Commission, Directorate-General Environment and produced a report that focusses on urban (buildings and industries) and agriculture. This report defines water efficiency as "the relationship between the amount of water required for a particular purpose and the amount of water used or delivered", which is a CE type concept.

Third, the Portuguese Water Use Efficiency Plan - PNUEA (National Laboratory of Civil Engineering 2001) is used in many activities related to water, such as, River Basin Management Plans (Portuguese Environment Agency 2019b), and Strategic Environmental Assessment of the Roadmap for Carbon Neutrality 2050 (Portuguese Environment Agency 2019a). PNUEA is defined for various types of water users, such as, urban, agriculture and industry. In its indicator section, it defines water use efficiency as the ratio of "useful consumption" to "effective demand", and sets water loss (%) equal to (100 – efficiency). This CE type equation does not explain the meaning of the words useful, consumption and effective. In practice, PNUEA uses Eq. (4.16) and defines various efficiency and investment targets that have been adopted by the decision-makers.

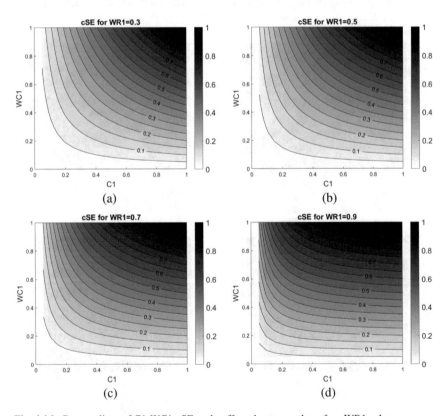

Fig. 4.16 Contour lines of C1-WC1-cSE trade-offs and patterns along four WR1 values

Finally, there are those who only use water quantity to calculate the performance of a system in the form of (water quantity/ total water quantity), which obviously is wrong. Section 6.2 gives an example with more explanation.

4.5.2 Water Productivity

WaP is defined as in Eq. (4.17) (Haie N., Sefficiency (Sustainable efficiency) of Water-Energy-Food Entangled Systems 2016):

$$WaP = \frac{production}{water\ quantity} \tag{4.17}$$

WaP does not give a percentage because production is different from water quantity, which mostly shows itself in the unit of the numerator being different from the denominator, e.g. €/mm. Some authors designate WaP as 'water use efficiency', and

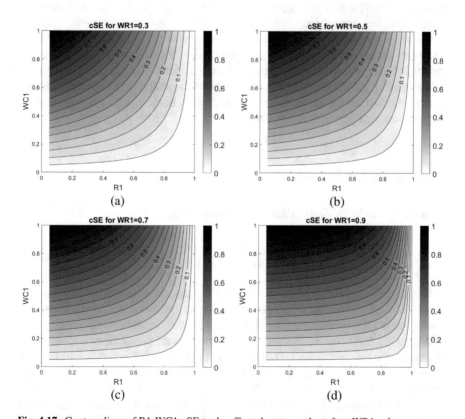

Fig. 4.17 Contour lines of R1-WC1-cSE trade-offs and patterns along four WR1 values

consequently much care should be exercised not to confuse its concept with efficiency indicators (in %) used in this book. Equation (4.17) applied to one specific
WUS has the following examples:

- 'production' can be yield (kg), mass of production (kg), monetary value (€),
  amount of product—different from water (m^3), etc.
- 'water quantity' can be water applied (VA in m^3), evapotranspiration (ET in mm),
  etc.

WaP as a ratio of output to input is similar to CE and equally a flawed indicator for
water management due to the issues given for CE, and various other reasons given
by some experts and organisations (Wichelns 2014; FAO & WWC 2015). There
are also variations to Eq. (4.17), such as 1/WaP and all are flawed. For example,
Coca Cola Company uses it under the name Water Efficiency meaning amount of
water used (litre) per amount of product made (litre) (Coca-Cola Company 2018). In
general, production depends on many inputs including water in a nonlinear fashion,
and as an input becomes scarcer, production becomes more dependent on that scarce
input. Under what combination of inputs, the productivity of the system (i.e., all
input considered) is good-enough? The answer to this question erroneously narrows
down to one (not many) input depending on the expert. For example, under the same
conditions for an irrigated agriculture, the answer of the water experts is proper
amount of water; the answer of the soil experts is better soil; the answer of the pest
experts is better pest control; the answer of the economic experts is about market or
land ownership, etc. Anyhow, Eq. (4.17) may prove to be valuable for agronomists and
particular industries but not for water managers who should aim for a comprehensive
and good-enough performance of a WUS.

4.5.3 Effective Efficiency

EE is defined as in Eq. (4.18) (Keller and Keller 1995):

$$EE = \frac{(ET - PP)_b}{\left(W_{qV1} * V1 - W_{qV2} * V2\right)} \tag{4.18}$$

EE is more complete than CE and meaningfully advanced the concept of water efficiency. However, it was not developed in a systemic and comprehensive manner and
consequently is an incomplete formulation and gives inaccurate results. Subtracting
PP (inflow) from ET (outflow) is not correct from the water perspective (and cannot be
applied to rain fed agriculture). There is no accounting for RP, which can be of great
importance, e.g., for groundwater. In addition, it does not include NR (a significant
flow in some applications) because EE is for irrigated lands only. On the other hand,
it does not comprehensively consider the Usefulness Criterion, i.e., water quality and
benefits. For example, the beneficial part of ET is in EE, but this distinction is not
extended to V1 and V2; and for quality, it only considers salt, i.e., $W_{qX} = 1 - LF_X$

(Sect. 1.3). Finally, although its name consists of the word "effective" but it was not defined, and as such, along the paper, it gets diverse meanings as applied to different things, such as effective inflow, effective precipitation, and effective efficiency.

4.5.4 Resiliency

RE is defined as in Eq. (4.19) (Loucks and Van Beek 2005; Hashimoto et al. 1982):

$$RE = \frac{\text{number of times a satisfactory value follows an unsatisfactory value}}{\text{number of times an unsatisfactory value occurred}}$$

(4.19)

Having in mind the definition of 'satisfactory' given in Sect. 2.4, Loucks and Van Beek (Water Resources Systems Planning and Management: An Introduction to Methods, Models and Applications 2005) define resilience as given in Eq. (4.19) stating that "Resilience can be expressed as the probability that if a system is in an unsatisfactory state, the next state will be satisfactory. It is the probability of having a satisfactory value in time period $t + 1$, given an unsatisfactory value in any time period t." Hence, resilience is an indicator of the response of the system, i.e., the speed of the recovery from an unsatisfactory condition (CDWR 2019). For example, a young person is more resilient than an old one, because she can recover faster from a sick (unsatisfactory) condition, such as a flu or Covid-19. In evaluating and improving resiliency, a system, such as a water network, has many resilience metrics that can be used depending on the scenario of interest. For an example in water networks, please refer to the WNTR software of U.S. Environmental Protection Agency (Klise 2017).

It is common knowledge that sustainable (water) systems must respond to social, economic and environmental dimensions of change. However, in many studies on WUS resiliency, the authors deal with one or two of the dimensions of sustainability or if all the three are used, they are done apart from each other, even though at the end they are, somehow, put together. In other words, there is a difference between a comprehensive integration of the three dimensions of sustainability (what we have in this book) and studying the three dimensions and then trying to integrate the results, usually partially. This is why the sustainable systems developed according to the theory presented in this book are also resilient, having in mind the following points:

- No system is absolutely sustainable or resilient, meaning that there are degrees to sustainability and resiliency. For example, a resilient system may respond well to a 50-year flood, but fails to recover under more sever ones.
- Any sustainable system must be resilient to foreseeable disruptions. Again, this is not absolute and it is possible to imagine sustainable systems that fail to a particular level of a specific disaster. No system can be highly resilient to all types of disruptions with all levels of intensity and extent.

- The relativity of these concepts does not mean that all systems are adequate. On the contrary, a sustainable and resilient system should be continually enhanced through learning (FIW5b; Sect. 3.2) in the context of improving Sefficiency in Sequity.

References

Ayers R, Westcot D (1994) Water quality for agriculture (FAO Irrigation and Drainage Paper 29), UN FAO (Food and Agriculture Organization), Rome, Italy

Benito P et al (2009) Study on water efficiency standards (final report). Bio Intelligence Service and Cranefield University, Paris

Berger A, Hill TP (2015) An introduction to Benford's law, 1st edn. Princeton University Press, sl

Brouwer C, Heibloem M (1986) Irrigation water management: irrigation water needs. UN FAO, Rome, Itay

CCME (2017) Canadian water quality guidelines for the protection of aquatic life: CCME Water Quality Index. [Online]. Available at: https://www.ccme.ca/files/Resources/water/water_quality/WQI%20Manual%20EN.pdf. Accessed 17 June 2019

CDWR (2019) Glossary—California Department of Water Resources (CDWR). [Online]. Available at: https://water.ca.gov/Water-Basics/Glossary. Accessed 23 Dec 2019

Coca-Cola Company (2018) Improving our water efficiency. [Online]. Available at: https://www.coca-colacompany.com/stories/setting-a-new-goal-for-water-efficiency. Accessed 9 July 2019

Coelli T, Rao D, O'Donnell C, Battese G (2005) An introduction to efficiency and productivity analysis, 2nd edn. Springer, Berlin

Dworak T et al (2007) EU water saving potential (Part 1 - Final Report). Ecologic Institute for International and European Environmental, Berlin

EEA (2015) EEA indicators. European Environment Agency, Copenhagen, Denmark

European Commission Water (2019) Building blocks: water efficiency activities (Last updated: 08-Jul-2019). [Online]. Available at: https://ec.europa.eu/environment/water/quantity/water_efficiency.htm. Accessed 28 Dec 2019

European Commission (2019) Environment. [Online] Available at: http://ec.europa.eu/environment/resource_efficiency/index_en.htm. Accessed 31 May 2019

European Parliament & Council (2000) Water Framework Directive, Official Journal L 327, European Union. [Online]. Available at: http://ec.europa.eu/environment/water/water-framework/index_en.html. Accessed 20 Mar 2018

FAO & WWC (2015) Towards a water and food secure future: critical perspectives for policy-makers. UN-FAO, Rome, Italy

Fitch E (2012) King midas: the seven deadly sins and externalities. Water Resour IMPACT 14(1):19

Haie N, Keller A (2012) Macro, meso, and micro-efficiencies in water resources management: a new framework using water balance. J Am Water Resour Assoc (JAWRA) 48(2):235–243

Haie N (2013) Sefficiency (sustainable efficiency): a systemic framework for advancing water security. Wuhan University, Wuhan and Yichang, Hubei

Haie N (2016) Sefficiency (sustainable efficiency) of water-energy-food entangled systems. Int J Water Resour Dev 32(5):721–737

Hashimoto T, Stedinger J, Loucks D (1982) Reliability, resiliency, and vulnerability criteria for water resource system performance evaluation. Water Resour Res 18(1):14–20

Hoekstra A, Chapagain A, Aldaya M, Mekonnen M (2011) The water footprint assessment manual: setting the global standard. Earthscan, London

Kahneman D, Rosenfield A, Gandhi L, Blaser T (2016) Noise: how to overcome the high, hidden cost of inconsistent decision making. Harvard Business Rev, October, pp 36–43

Keller A, Keller J (1995) Effective efficiency: a water use efficiency concept for allocating freshwater resources. Winrock International, Arlington, Virginia

Klise KA et al (2017) Water Network Tool for Resilience (WNTR) user manual. U.S. Environmental Protection Agency (EPA/600/R-17/264), Washington, DC

Loucks D, Van Beek E (2005) Water resources systems planning and management: an introduction to methods, models and applications. UNESCO, Paris

Loucks DP (2002) Quantifying system sustainability using multiple risk criteria. s.l., Cambridge University Press

MathWorks (2018) MatLab. [Online]. Available at: MathWorks MatLab https://www.mathworks.com. Accessed 19 Mar 2018

Maxwell D et al (2011) Addressing the Rebound Effect. European Commission DG Environment, sl

MEEC (1997) Environmental quality standards for surface water. [Online]. Available at: http://english.mee.gov.cn/SOE/soechina1997/water/standard.htm. Accessed 17 June 2019

MEEC (2018) 2017 Report on the State of the Environment in China. [Online]. Available at: http://english.mee.gov.cn/Resources/Reports/soe. Accessed 19 June 2019

Merriam-Webster Dictionary (2019) Target. [Online]. Available at: https://www.merriam-webster.com/dictionary/target. Accessed 31 May 2019

NASDAQ-100 n.d. NASDAQ-100 equal weighted index shares (QQQE). [Online] Available at: https://www.nasdaq.com/symbol/qqqe. Accessed 23 June 2019

National Laboratory of Civil Engineering (2001) Programa Nacional para o Uso Eficiente da Água. National Institute of Water (INAG), Lisbon

Nigrini M, Miller S (2007) Benford's law applied to hydrology data: results and relevance to other geophysical data. Math Geol 39(5):469–490

OECD (2001) Water quality index. [Online]. Available at: http://stats.oecd.org/glossary. Accessed 30 June 2012

Oxford Dictionary (2018) 'meso-'. [Online]. Available at: https://en.oxforddictionaries.com/definition/us/meso. Accessed 19 Mar 2018

Portuguese Environment Agency (2019a) River basin management plans—3rd cycle. Portuguese Environment Agency, Lisbon

Portuguese Environment Agency (2019b) Roadmap for carbonic neutrality 2050/strategic environmental assessment. Portuguese Environment Agency, Lisbon

Rawls J (1999) A theory of justice, Revised edn. Harvard University Press, Cambridge

Seckler D, Molden D, Sakthivadivel R (2003) The concept of efficiency in water-resources management and policy. s.l., CABI Publishing and International Water Management Institute

Sen A (1992) Inequality reexamined. Oxford University Press, New York

Stone D (2012) Policy paradox: the art of political decision making, 3rd edn. W. W. Norton & Company Inc., New York

UNEP (2007) Global drinking water quality index development and sensitivity analysis report. GEMS (Global Environment Monitoring System) Water Programme Office, Burlington, Ontario, Canada

UN-HLPW (2017) Bellagio principles on valuing water. United Nations, Sustainable Development Goals (SDG), High Level Panel on Water. [Online]. Available at: https://sustainabledevelopment.un.org/content/documents/15591Bellagio_principles_on_valuing_water_final_version_in_word.pdf. Accessed 29 Sept 2017

Weisstein EW (2018) Benford's law. MathWorld, A Wolfram Web Resource. [Online]. Available at: http://mathworld.wolfram.com/BenfordsLaw.html. Accessed 24 Mar 2018

Wichelns D (2014) Water productivity: Not a helpful indicator of farm-level optimization. [Online]. Available at: http://www.globalwaterforum.org/2014/11/11/water-productivity-not-a-helpful-indicator-of-farm-level-optimization. Accessed 2 July 2019

Willardson LS, Allen R, Frederiksen H (1994) Elimination of irrigation efficiencies. USCID, Denver, Colorado

Chapter 5
Sequity (Sustainable Equity)

The idea of equity is related to development, which is "a process of expanding the real freedoms that people enjoy." (Sen, Development as Freedom 1999) Furthermore, "Freedom can be distinguished *both* from the *means* that sustain it and from the *achievements* that it sustains." (Sen, Inequality Reexamined 1992) In this book, we focus on water as a means with the understanding that it is not a recipe for "equality of well-being achieved" (Sen, Inequality Reexamined 1992), nor a complete implementation of social justice theories presented by distinguished people, such as, Rawls (A Theory of Justice 1999) and Sen (The Idea of Justice 2009).

For the purposes of this book, sustainable water equity "means a fair share considering the water needs and the ability to use the water efficiently" (Dellapenna 2001). These two requirements, i.e., need and efficiency, have degrees (i.e., they are not binary) and correspond to two principles of (water) equity: distributive and aggregative, respectively (FIW5c). Many authors and entities in trying to define and explain sustainable water use stress the two requirements or principles (Loucks and Van Beek 2005; International Law Association 2004). Before continuing with these two, it is highly significant that they should operate in an environment of fairness, as the above quote emphasizes. Here, it is sufficient to reiterate that the concepts and procedures set forth in this book are in accordance with fairness, which promotes transparency (FIW1), reasonable actions (FIW5) and impartiality (FIW1 and FIW5).

The distributive principle is about *members*, such as individuals, stakeholders, groups and zones, and their 'need', which is defined as the "essential or very important, not just because you would like to have them" (Oxford Dictionary 2019). This definition conveys some kind of conditionality, meaning that (water) 'need' must be above a minimum as will be developed in the next subsection.

Chapter 4 is about systems as a whole, more specifically WUSs and their aggregative behaviour, which is presented in this book as Sefficiency. The focus of the current chapter is about the performance of a system from the point of view of both the

N. Haie, *Transparent Water Management Theory*, Water Resources Development and Management, https://doi.org/10.1007/978-981-15-6284-6_5

aggregative and distributive principles, which are explicitly revealed by focussing on distributive equity and Sefficiency. Similar to this arrangement is the Difference Principle of Rawls (A Theory of Justice 1999) as explained by Sen (The Idea of Justice 2009): "The second part of the second principle (called the 'Difference Principle') is concerned with distributive equity as well as overall efficiency, and it takes the form of making the worst-off members of the society as well off as possible." Let us remember that the distributive-aggregative dichotomy from a water-centric point of view is part of the theory elaborated earlier in this book. For instance, this implies that the state of institutions and technologies are not dealt with directly in this Chapter, but both are bounded by the FIWs in the context of a learning approach discussed in Chap. 3.

The rest of this chapter develops an approach to enable transparent communication, to focus attention on problematic situations through targeting, and to elaborate policy guidelines in order to coherently and consistently solving those management issues.

5.1 Categories

The interactions between water managers, politicians and people are sometimes intense and more so as water scarcity, global warming and problematic precipitation patterns increase. How, then, water managers can present their policies in such a way that reveal elegance, insight, and strategy, while being aligned to societal objectives? For the purposes of this book, the answer is through water categories by setting fair thresholds. "Political reasoning involves metaphor-making and category-making, but not just for beauty's sake or for insight's sake. It is strategic portrayal for persuasion's sake and, ultimately, for policy's sake."(Stone 2012) Transparent categories make the art of persuading people and politicians much easier. For example, explaining the percentage of a city population under the category 'Extreme Water Poverty' significantly facilitates the reason for investing in water infrastructures for that group.

For these reasons, employing categories is a routine practice. Few examples are:

- Water Framework Directive of the European Union (European Parliament & Council 2000) with five classes and corresponding colours for the ecological status of surface waters
- Human Development Index with four classes (UNDP-HDR 2018)
- Climate Change Performance Index with five categories from "very high" to "very low" (Burck et al. 2018)
- United Nations (UN-SDG 2019) with five categories about the proportion of the global population that uses various levels of drinking water services

There are many ways defining categories that cover all the possibilities of an indicator or descriptor. In this book, we consider the amount of water allocated to a member in a period of time as the *indicator of interest* (refer to Sect. 2.3 for explaining 'allocation'). Employing four or five categories is common, with five being more insightful and easy to summarize in three categories, and hence the option in this book. To set up this choice for the indicator of interest, we need four thresholds, which are the boundaries between two consecutive categories. Sometimes the naming, their four values and corresponding colours are already defined within laws and regulations, e.g., at the national level, however we employ the following generic naming with Amax > Max > Min > Amin:

- Absolute minimum (Amin): the value of the indicator of interest for any member must be greater than or equal to Amin
- Minimum (Min): the value of the indicator of interest for any member must be greater than or equal to Min under non-transitory conditions
- Maximum (Max): the value of the indicator of interest for any member must be less than or equal to Max under non-transitory conditions
- Absolute maximum (Amax): the value of the indicator of interest for any member must be less than or equal to Amax

Transitory conditions should be rare, and the system design should be under normal or non-transitory state. Let us be clear that these thresholds are set at the lowest level of authority as possible. For example, if there is no national level regulation on any of these thresholds, then local water managers should define them in consultation with all the stakeholders. Anyhow, in setting the thresholds, international guidelines such as the following can be very helpful:

"One of the most important recent milestones has been the recognition in July 2010 by the United Nations General Assembly of the human right to water and sanitation. The Assembly recognized the right of every human being to have access to sufficient water for personal and domestic uses (between 50 and 100 litres of water per person per day), which must be safe, acceptable and affordable; (water costs should not exceed 3 per cent of household income), and physically accessible (the water source has to be within 1000 m of the home and collection time should not exceed 30 min)." (UN-Water 2019)

The Amin and Min thresholds are rather common in programs and publications, however, the Amax and Max thresholds are less discussed and analysed. Almost everything in our universe work according to an equilibrium between minimum and maximum for most of descriptors, because outside these extremes the system becomes unsustainable. Let us see examples:

- Why cells of our body should stop dividing after their maximum? (if not, we call it cancer)
- Why human body temperature has maximum?
- Why speed in the physical universe has a maximum? (of light)
- Why plants have maximum rate of transpiration?
- Why speed of the automobiles in a highway should have maximum?
- Why the discharge of a specific water pump has a maximum?

The answer to these and many other similar questions is that we live in a vast and deep interconnected world, and the necessary human and nonhuman relationships proceed from the reality of things, and not our superficial and superstitious perceptions (at least in long term). This has many meanings, including the fact that there are limits and conditions around everything (physical, emotional, intellectual, spiritual – individual, societal, civilizational…), even though they change in space-time continuum. Consequently, maximums are everywhere and are fundamental requirements for sustainable development and certainly for utilizing water, particularly under water scarcity. In reality, we should become transparent about such a limit also because it has been (will be) forced upon us: "Physical, economic, and ecological limits constrain the development of new supplies and additional water withdrawals, even in regions not previously thought vulnerable to water constraints. New kinds of limits are forcing water managers and policy makers to rethink previous assumptions about population, technology, regional planning, and forms of development" (Gleick 2010).

Maximums are dynamic in the sense that it may change in time. For example, after five years, the water managers of a region may have enough data that can lead them to decrease the maximum or reflect and predict with the stakeholders that the maximum should be decreased, if water scarcity continues to increase. Doing so, the categories of the members of a WUS may change, leading to new challenges.

Maximums should not be seen as rationing, which is needed for emergency or transitory situations and, generally, the public does not like it. Furthermore, there are those that advocate water pricing to achieve sustainability, but without physically controlling water by putting maximums, it proves to be very difficult (Dinar et al. 2015). It should be noted that effective water pricing has its own merit (FIW5a), however in many cases such a policy has impacts that are more devastating to the least advantaged and goes against equity and fairness. Additionally, almost everything depends on water and with price increase, the price of other things also have a tendency to increase, which again affects the least advantaged.

Consequently, a Water Use System and its members should obey minimums and maximums. It is one of the main duties of water managers (within a learning approach) to set these boundaries to produce adequate and reasonable functioning. Having done so, equity demands comparison of at least two members regarding the indicator of interest already defined in this subsection. Let us designate these two members as M1 and M2 to refer to the Most and Least Advantaged Members, respectively. This type of differential focuses our attention on the extremes, which are less sustainable, while presenting the path for a more balanced and sustainable combination. Showing these two members along two axes in conjunction with the thresholds just defined, a general and valuable structure emerges as drawn in Fig. 5.1. It is important to note that the space of all the possible feasible systems is below the $M1 = M2$ line because of our designation that $M1 \geq M2$.

As mentioned, the four thresholds produce five categories shown in Fig. 5.1 as five rows (along M2) and five columns (along M1) numbered 1 to 5 according to the labels given in Table 5.1.

The $M1 = M2$ line in Fig. 5.1 represents numerical equality in water allocation between the two members. As our systems operate to the right of this line, inequality increases, but may or may not have Water Poverty because:

- Point A gives $M1 = M2 = Amin$

- Point B gives M1 = M2 = Min
- Point C gives M1 = M2 = Max
- Point D gives M1 = M2 = Amax

Although the impacts indicated in Table 5.1 are self-explanatory, they will become clearer as we go forward in this chapter. For example, many water systems around the world sometimes operate under Extreme Water Poverty to the extent that people die, directly or indirectly, due to the lack of proper freshwater. By indirectly, for example, we mean hunger because water is indispensable to food production.

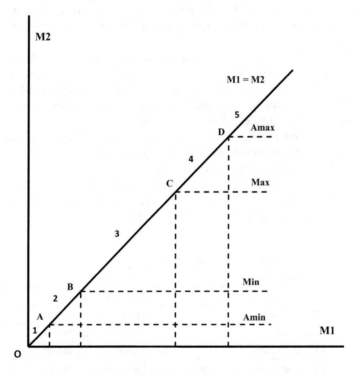

Fig. 5.1 A general structure for comparing categories of a system

Table 5.1 Water categories, codes and impacts

Category code	Water category	Impact
1	Extreme Water Poverty	Death (avoid totally)
2	Water Poverty	Survival (transitional)
3	Water Wise	Equitable
4	Water Abundance	Unsustainable (transitional)
5	Extreme Water Abundance	Insecure (avoid totally)

5.2 Segments

Figure 5.1 has 15 segments, which are about water allocations, and sometimes with correspondence to classes of population, such as poor, marginalized, or wealthy people. Here, we define a segment as all the possible water arrangements that fall within the union of a category of M1 and another of M2. Consequently, segmentation dramatically reduces the high diversity inherent in all the possible combinations of M1 and M2 by focussing our attention in one segment. However, there are still differences between various systems of a segment but they are within the confines of purposely-defined thresholds with specific societal objectives. Of course, water managers can have further division of a segment if it improves transparency and fair decision-making.

Segment (Sg) notation is by the row (along M2) and column (along M1) using the category codes 1 to 5 indicated in Fig. 5.1. For example, Sg23 is the rectangle with point B as its upper left vertex, i.e., M1 in Water Wise category and M2 in Water Poverty. Table 5.2 shows the 15 segments and their categories, and Table 5.3 gives the distribution of the categories for each member.

Table 5.3 gives the inherent distribution of all levels of allocation without regulation. It shows that the share of water poverty (= Extreme Water Poverty + Water Poverty) and water abundance (= Extreme Water Abundance + Water Abundance) for M1 and M2 is inversed, meaning that water poverty for M1 is 20%, and for M2 60%, while water abundance is 60% for M1 and 20% for M2. Consequently, to promote Water Wise systems, i.e., achieving the most equitable and sustainable

Table 5.2 Segments and their corresponding water categories

Sg	M1 category	M2 category
11	Extreme Water Poverty	Extreme Water Poverty
12	Water Poverty	Extreme Water Poverty
13	Water Wise	Extreme Water Poverty
14	Water Abundance	Extreme Water Poverty
15	Extreme Water Abundance	Extreme Water Poverty
22	Water Poverty	Water Poverty
23	Water Wise	Water Poverty
24	Water Abundance	Water Poverty
25	Extreme Water Abundance	Water Poverty
33	Water Wise	Water Wise
34	Water Abundance	Water Wise
35	Extreme Water Abundance	Water Wise
44	Water Abundance	Water Abundance
45	Extreme Water Abundance	Water Abundance
55	Extreme Water Abundance	Extreme Water Abundance

Table 5.3 Water categories and their segment distributions in M1 and M2

Water category	% M1	% M2
Extreme Water Poverty	6.7	33.3
Water Poverty	13.3	26.7
Water Wise	20.0	20.0
Water Abundance	26.7	13.3
Extreme Water Abundance	33.3	6.7

WUS, we need concerted effort in managing water in the direction of eradicating the other four water categories.

Increasingly, the main focus of water managers is in bringing M2 out of water poverty with little attention to water abundance of M1, which is 60% under the inherent pattern of categorization presented in this Chapter. Hence, the problems manifest themselves in two ways: to reduce M2 water poverty as well as M1 water abundance, each at 60%. This Chapter, through the theory presented in this book for water management, promotes Equality to eradicate water poverty, Conservation to decrease the use of various resources (water quantity, water quality, energy, land, money, etc.), and Sefficiency to better advance Water Wise development. These three parts of equity will be discussed more fully in the next subsections.

5.3 Equity Revisited

Equity as fairness operates under three distinct principles, namely equality, conservation and efficiency (mostly in this order). Consequently, it puts to rest the arguments of those who object to any one of these three water principles. If each one of them can be rejected through (valid) reasoning, they together in the context of a learning approach presented in this book may prove to be difficult, or rather impossible, to set aside. Let us summarize each of these three.

5.3.1 Equality

The principle of equality ensures the existence of necessary and sufficient conditions in order to eliminate poverty. This principle guarantees a 'minimum' allocation to all concerned, hence promoting the objectives of the society and the idea of the UN human right to water (UNGA 2010). In other words, equality should promote water (re)allocation in order to make water poverty non-existent, meaning that no member should be below the Min threshold as shown in Fig. 5.1, eliminating the categories of Water Poverty and Extreme Water Poverty. In this context, the driving force for (re)allocation is the numerical concept of equality, meaning that all the effort should go into advancing toward and finally achieving M1 = M2. After establishing this equality, all the water arrangements at least have equality as defined at the beginning of this paragraph. Let us see in terms of segments:

- Sg34 has equality, because none of its systems operates below the Min threshold.
- Sg11 must, at least, establish numerical equality at Amin threshold.
- Sg13 must go toward equality by water reallocation in order to get to Sg23.
- Sg23 achieves equality in reaching Sg33.
- Sg12 should go toward equality in reaching Sg22.

Equality does not mean uniformity. Actually it is impossible to produce such a condition of uniformity because of the enormous diversity in human behaviour, characteristics and environment. However, "The Pigou-Dalton principle of transfers says that inequality decreases (or social welfare increases) when an even transfer is made from a richer to a poorer individual without reversing their pairwise ranking" (Fleurbaey 2016). For more information on this principle, the reader can also refer to Sen (1992). This principle also applies to water, which particularly presents higher positive impacts for the segments below the Min threshold. It should be noted that in this book there are three meanings in transferring water. First is allocation, which is transferring new water from a source to a member, second is about reallocation that can happen from M1 to M2, and the third is stopping allocation, i.e., ending the abstraction of water from a source and leaving it in nature. If a segment is operating under the condition of numerical equality, to improve its situation the policy must maintain equality while moving towards better Targets, e.g., from A to B.

Finally, there is the objection that "a policy of attainment equality would lead to a very 'low level equality' for all" (Sen, 1992). Then he goes on to state that there is "some force in that objection, but it would have been a more telling counterargument had equality been the only principle to be used." And this is the situation in the theory presented in this book, which includes the principle of equality. Additionally, water management must consider other laws (e.g., the law of water balance) and principles (e.g., the principle of the water conservation) and apply them all together in a coherent, consistent and efficient manner.

5.3.2 Conservation

Conservation is "for avoiding the unnecessary use of natural materials such as wood, water, or fuel" (Cambridge Dictionary 2019a), hence essential for sustainable development. In this definition, "unnecessary" is the keyword and as related to water equity in this chapter, conservation promotes reallocation in order to make water abundance non-existent. This means that no member should operate above the Max threshold, avoiding the categories of Water Abundance and Extreme Water Abundance.

Indeed achieving equality is more urgent than implementing conservation. However, in many situations, particularly those under water scarcity, the implementation of equality depends on conservation measurements in order to release water to a member under water poverty. This conservation action is priority for at least four segments, namely Sg14, Sg15, Sg24 and Sg25.

Even under the conditions that water poverty does not exist, conservation is still necessary. In segments Sg34, Sg35, Sg44, Sg45 and Sg55 conservation leaves more water with better quality in nature, and promotes the use of less energy, land and money. Consequently, water conservation minimizes harm to water and many other resources.

Finally, water maximum thresholds, rationing, conservation and saving are all different. While maximum thresholds may promote conservation, the idea of water conservation is broader and is not about maximums, rationing or saving. On the other hand, water conservation may promote water saving but their relationship suffers a myth (Sect. 6.1). Rationing occurs under transitory, i.e., temporary conditions in contrast to the other three words.

5.3.3 Sefficiency

After removing water poverty and water excesses, what remains is Sg33, meaning that both M1 and M2 are operating in the Water Wise category. In this segment, equality can be numerically established (i.e., along the line BC), however "the worst form of inequality is to try to make unequal things equal" (Aristotle, 384 BC-322 BC). Consequently, equity brings fairness in diversity by using the Sefficiency framework developed in Chap. 4. The best point to operate within the Sg33 segment can be found by using the full Sefficiency, which depends on the amount of available water, its quality and benefits. Furthermore, sometimes it is necessary to apply Sefficiency to Sg13 and Sg23 as explained in Sect. 5.5 below.

We know that there may be various Sefficient scenarios for a WUS, but not all of them are fair. In this regard, equity developed in this Chapter is necessary to single out one Sefficient distribution or dramatically reduce the number of scenarios for fair decision-making in accordance with bounded rationality introduced in Chap. 1. In other words, there is an inherent priority among Equality, Conservation and Sefficiency, with the first two being constraints on the third. There are other limiting factors, such as, water right (different from UN human right to water) that should be considered in its due place in equity, e.g., in the IN/OUT flows of a WUS.

As a final note on Sequity, we should mention that in many circumstances such as for public information, colour works better than numbers or symbols. Using three colours, Table 5.4 presents an example.

The Sefficiency values, particularly the yellow-green frontier of 80%, depend on sectors and types of Sefficiency (e.g., $ic = 0$ or 1) and some of the examples in Chap. 6 demonstrate this.

Table 5.4 Equity limits in three colours

Equity	Red	Yellow	Green
Equality	M2 < Amin	M2 < Min	M2 ≥ Min
Conservation	M1 > Amax	M1 > Max	M1 ≤ Max
Sefficiency (%)	<60	<80	≥80

5.4 Targets

For establishing equity, there is the necessity of defining targets. "Setting targets establishes concrete quantifiable objectives, while focusing attention on the issues at hand and providing incentives to take action and mobilize the resources necessary for reaching the goals" (UNESCO-WWAP 2006). It should be added that to properly understand the complexities of progressing toward the targets, "regular and reliable monitoring is required" (UNESCO-WWAP 2006). Furthermore, the learning approach briefly presented in Sect. 3.2 needs the information of such a monitoring system, but at the same time it is required for data gathering and sustainable development.

In order to simplify the complex consultation and decision making associated with targets and priority (re)allocations, we chose the thresholds to also play the all-important role of targets. However, in avoiding confusion between thresholds and targets, we employ the letters that correspond to the thresholds as shown in Fig. 5.1, which are A, B and C (D, i.e., Amax should not be a target). For clarity, let us present their definitions:

- Target A: the lowest category among the members is Extreme Water Poverty, which water (re)allocation must eliminate with alarming urgency.
- Target B: the lowest category among the members is Water Poverty, which water (re)allocation is to eliminate with urgency.
- Target C: triggering this target promotes the application of the Conservation principle.

Targets for M1 and M2 are specified for each segment in Table 5.5 with the priority targets to focus the primary (re)allocation operations. For example, in Sg12, the target for M2 is A, which is the priority target of reaching to the Min threshold, and the non-priority target for M1 varies between A and B: [A, B] = A ≤ Target M1 ≤ B. Furthermore, along rows 1 and 2, there should be the understanding that the water situation of M2 can never be worsen.

Table 5.5 Targets and priorities for each segment and member

Sg	M1 Target	M2 Target
11	A	A
12	[A, B]	A
13	[B, C]	A
14	[B, C]	A
15	[B, C]	A
22	B	B
23	[B, C]	B
24	[B, C]	B
25	[B, C]	B
33	[B, C]	[B, C]
34	C	[B, C]
35	C	[B, C]
44	C	C
45	C	C
55	C	C

Note: darker shade = Priority, lighter shade = Sefficiency

5.5   Four Policy Types

After defining targets and priorities, some aspects of (re)allocation policies are discussed here. However, let us define the following new variables important in continuing this Chapter:

- n: 1 or 2 in Mn, i.e., M1 or M2; M2 ≤ M1
- Tgn: Target (Tg) of M1 or M2; Tgn > 0
- SW: Water Shortage, i.e., the amount of water needed to reach target; SW ≥ 0
- SWn: SW of Mn from its Target; = Tgn – Mn; SW1 ≤ SW2
- ZW: Water excess, i.e., the amount of water in excess of target; ZW ≥ 0
- ZWn: ZW of Mn from its Tg; = Mn – Tgn
- NEW: New water supply; 0 ≤ NEW ≤ (SW1 + SW2)

Any system at a particular time is defined by a point on one of the 15 segments shown in Fig. 5.1. For example, G in Fig. 5.2 is such a system, which exhibits water shortages with their targets. In general, water (re)allocation policies must deal with shortages, excesses and water from new sources. These three factors describe the (re)allocation needs of segments and can be summarized into four policy types (Pt) as presented in Table 5.6.

Let us see some of the characteristics of the four policy types, having in mind that the overriding policy for a WUS is to eradicate the two categories of Extreme Water Poverty and Water Poverty.

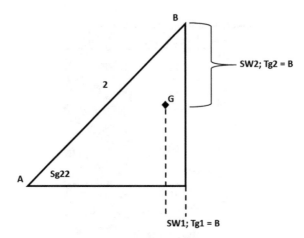

Fig. 5.2 The direction of achieving equality in a triangular segment

Table 5.6 Policy types, segments, (re)allocation needs and equity focus

Policy type	Sg (Segment)	(re)allocation needs	Equity focus
PtI	11, 12, 22	SW2, SW1, NEW	Equality
PtII	13, 14, 15, 23, 24, 25	SW2, ZW1, NEW	Equality
PtIII	34, 35, 44, 45, 55	ZW1, ZW2	Conservation
PtIV	33	–	Sefficiency

Policy Type I

Policy type I (PtI) covers three segments, namely Sg11, Sg12, Sg22, and manifests only water shortage, which needs new water to fulfil its allocation requirements, i.e., eliminating the two categories of poverty. In order to reach the targets, NEW water (i.e., water not in use) by the amount of SW1 + SW2 must be urgently found and allocated. There are serious problems for the three segments and an unwavering effort is needed to get out of them by mostly going toward numerical equality, i.e., M1 = M2. This is a relatively easier task for Sg11 and Sg22 (the two problematic triangular segments) than for Sg12, which puts M2 in Extreme Water Poverty category and makes it the priority member (Table 1.5).

The policies depend on how fast NEW is available, how many people are affected in each of M1 and M2, and the force of the voices in the stakeholder meetings and among the decision makers. However, there are situations that common sense decision-making is easier. For example, let us examine the water arrangements near the four corners of a rectangular segment, such as Sg12, with NEW = 0.

- If the water allocations are near the lower right corner of the rectangle, then reallocation from M1 to M2 should be top priority by going toward equality M1 = M2 = A.

- In the upper right corner, M2 can easily reach A but M1 gets a little more distant from B.
- In the upper left corner, both must go toward and probably reach M1 = M2 = A.
- In the lower left corner, the situation is extremely urgent and new water should be allocated to M2 with the target being M1 = M2 = A.

It should be noted that M1 does not have excess water (ZW) but it has more than M2 that under this condition can help it to come out of the Extreme Water Poverty or get closer to the equality line. In other words, if M2 is below A, there is an alarming urgency to take it out of this category without lowering M1 to such a disastrous category.

Policy Type II

Policy type II (PtII) covers six segments, namely Sg13, Sg14, Sg15, Sg23, Sg24, Sg25, and manifests both water shortage, and water excess for reallocation. Segments 14, 15, 24 and 25 are more straight forward as the excess water of M1 can be reallocated to M2, having in mind the M1 targets of [B, C]. As an example, let us suppose that a WUS is operating in Sg15, which puts it in Target A with sever water shortage for M2 (SW2) and much excess water from M1 (ZW1). If we plan to reallocate some of the ZW1 to M2 in order to reach A (i.e., the Amin threshold), then the system say gets to Sg24. This is because we reduced water allocation to M1, hence changing it from Extreme Water Abundance to Water Abundance. However, M2 is in Water Poverty and needs more water to move to Water Wise (Sg33), which can be done by further reallocation of water from M1 to M2, with M2 operating at B (the Min threshold) and M1 somewhere between B and C. Although this reallocation example takes the WUS out of water poverty, we need to move forward and check for the possibilities of making the system Sefficient, using the ideas of Policy type IV given below. For these four segments of PtII, if ZW1 cannot satisfy SW2, then new water is needed, which take us to the following procedure.

The points mentioned in the rest of this subsection apply to all the segments of PtII, however let us focus on the more complex segments, i.e., Sg13 and Sg23. One clarification is worth mentioning considering Table 1.5, which gives a range for the target of a member in Water Wise category. However, for Sg13 and Sg23, we should use B as the target for M2 because of the (Extreme) Water Poverty situation of M1. In other words, ZW1 is defined relative to B and as such presenting the potential amount of water for reallocation, but it does not mean that M1 must be reduced to B. To continue, we define a fraction called Reaf in Eq. (5.1):

$$Reaf = \frac{SW2 - NEW}{ZW1}, 0 \leq Reaf \leq 1, ZW1 > 0 \tag{5.1}$$

Where Reaf is the Reallocation fraction, i.e., fraction of ZW1 that satisfies SW2 after considering NEW. Some of the complexity in water (re)allocation comes from Reaf because some stakeholders (e.g., M1) want it to be closer to zero, and others (e.g., M2) prefer it to be meaningfully above zero, particularly in the absence of

Table 5.7 Policy possibilities for Segments 13 and 23

Start Sg	Condition	M2	M1	End Sg
Sg13 i	Reaf $= 0$	To A	–	Sg23
ii	$0 < $ Reaf $ \leq 1$	To A	To/Toward B	Sg23
iii	NEW $+$ Reaf $*$ ZW1 $-$ SW2 < 0	Toward A	To/Toward B	Sg13
Sg23 i	Reaf $= 0$	To B	–	Sg33
ii	$0 < $ Reaf $ \leq 1$	To B	To/Toward B	Sg33
iii	NEW $+$ Reaf $*$ ZW1 $-$ SW2 < 0	Toward B	To/Toward B	Sg23

NEW. However, if ZW1 is zero (i.e., M1 = B), then NEW should be allocated to M2 to satisfy its SW2. Besides this special case, Table 5.7 presents these two segments and their policy possibilities.

Each segment has three starting points designated as i, ii and iii with the first two having enough water for allocation and/or reallocation to satisfy SW2.

i. The first one (i) can happen when there is enough NEW to fulfil SW2 regardless of ZW1, i.e.., Reaf $= 0$, which makes M2 to reach its target and get to a higher category.

ii. The second (ii) can happen if Reaf varies between zero and one, including one, meaning that a percentage of ZW1 (with or without NEW) satisfies SW2. Under this condition, water allocation to M1 goes toward B (may decrease to B), while M2 reaches its target and enters a higher category. The major problem here is for the water managers to decide on a meaningful Reaf.

iii. Under this situation, SW2 is not satisfied, and consequently M2 does not reach its target and remains in its starting negative category. This happens where there is not enough NEW, and evenif water managers decide to divert all of ZW1 to M2 (i.e., Reaf $= 1$), which translates to no more internal water for reallocation (unless lowering M1 target to A – a very hard task).

Finally, it should be noted that the amount of water in a (main) source after abstraction may be unhealthy, i.e., it is below the minimum required for a properly functioning ecosystem. This can happen even though water quantity before abstraction (VU) and after return (VD) are adequate, as shown in an application on urban equity in the next Chapter.

Policy Type III

Policy type III (PtIII) covers five segments, namely 34, 35, 44, 45, 55 and manifests water excess in the amount of ZW1 $+$ ZW2, which should not be abstracted from water bodies as a conservation measure. For example, Sg34 has ZW2 $= 0$ with ZW1 $=$ M1 $-$ C, and Sg45 has excess water equal to ZW1 $+$ ZW2 $=$ M1 $+$ M2 $-$ 2C.

Policy Type IV

Policy type IV (PtIV) should deal with Sg33, and Sefficiency must be applied for sustainable (re)allocation of water (refer to Chapter 4 for more details). There are also other situations that only one member is in Water Wise category (Sg 13, 23, 34, 35) and may prove to be tempting to apply Sefficiency to these segments too. However, the principles of equality and conservation have precedence over efficiency. In this context, Sg34 and Sg35 must become Sg33 through conservation measures before applying Sefficiency. Sg13 and Sg23 need to go to Sg33 before applying Sefficiency, however, it is possible that through the application of Sefficiency, more water becomes available for reallocation, hence facilitating the advancement of these two segments to Sg33.

5.6 Reality Check

In looking into the policies and ideas presented in this book, two real issues worth briefly discussing.

First that water is life and as scarcity increases, water power goes further 'upstream', meaning that influences of the privileged and decision makers increase, making the implementation of some of the ideas in this book more difficult. For example, let us do a reality check for Sg11 and Sg22:

- Sg11 most likely becomes Sg13 than Sg22
- Sg22 most likely becomes Sg24 than Sg33

Besides these real tendencies, the hope is that, as water managers understand the fundamentals presented in this book, such as the terminologies, FIW1 and FIW2 of the theory, the need of a regular learning approach, and the interconnectedness of the three principles presented in equity, their decisions will advance in the right direction as years pass by. This indeed will be beneficial for both M2 and M1 because water interrelates many crucial aspects of the collective life.

Second, water managers sometimes face with existing water structures that are designed to function for decades, and consequently, the idea goes that water (re)allocation to new places (M2 areas) is not practical and it is costly. Under such conditions or situations of uncertainty, system management and design should follow adaptive procedures, which include a learning process (see Sect. 3.2) having in mind the following two points:

- The existence of infrastructures should not be the reason for perpetuating inequality and wasting water. The idea of water equity consisting of equality, conservation and Sefficiency has priority even if the result is to utilize part of an infrastructure.
- Lack of sufficient capital for funding more infrastructure to advance equality is of importance. However, national, regional and global programs, such as the programs of the United Nations can help very much. One thing should be clear that more than funds is the existence of real transparent plans to solve these issues. This means that, for example, if a member is under water poverty and after ten years nothing improves, or probably worsens, then this is not acceptable nor ethical.

We should have no doubts that donors and investments increase if a transparent and equity driven policy (as given in this book) presented to them.

Finally, let there be no mistake of the intentions of this book. The vision is not to create a theory and a corresponding approach that overnight solves the problems of water management all over the world! On the contrary, and similar to all things in life, it is about threading a path according to the ideas presented in this book, which over time can make water systems more sustainable and equitable. The forces of reality and human relationships are much more complex and need many fundamental changes during years and decades, even if we thread a sustainable path in water (Appendix C: Environment and Social Contract). In this context, I encourage water managers to focus, at least, on implementing parts of this book, such as, making explicit progress in time with the concepts presented in this Chapter.

5.7   Phases in Decision-Making

Sequity is a significant tool in achieving a good-enough solution along the path of the theory developed here. Water management is highly complex with fuzzy and noisy actors and data. The process of decision-making, i.e., coming up with a unique solution/choice for implementation needs many judgment calls (see Sect. 1.2). Judgment means "the ability to make decisions or to make good decisions" (Cambridge Dictionary 2019b) and whether we make "decisions" or "good decisions" largely depends on using Sequity, because of its relative objectivity that substantially reduces noise (Sect. 4.3).

Figure 5.3 shows the water management phases, steps and feedbacks to come up with a solution for a WUS. The labels of the four phases are System, Accounting, Sequity and Solution, which follow the FIWs of the theory presented in Chapter 3. There are three steps in each phase with the possibility of feedback, i.e., going back to an earlier step, as the learning process advances. Of course, every step can utilize various models and methods for better characterisation, which can be found in many books and websites.

The System phase (I) is to become transparent about the scope of the WUS, stakeholders and objectives. It is crucial to draw the schematic of the WUS (e.g., Figure 2.1), including the identification of the main source of water, and list the initial objectives and stakeholders. Both of these, usually, change as the process of action, reflection and consultation (Sect. 3.2) develop along various phases and steps. One thing is certain along these phases and that is the increase of stakeholders, particularly if there exists water scarcity.

The Accounting phase (II) is about data and quantifying the three Pillars of water management with the active involvement of the stakeholders. These data are descriptive in nature and mostly noisy (Sect. 4.3), which should be dealt with accordingly. Besides what has been written so far about the three Pillars, let us make some general comments for the sake of this phase.

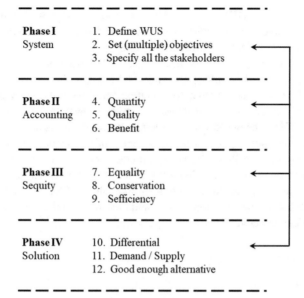

Phase I System	1. Define WUS 2. Set (multiple) objectives 3. Specify all the stakeholders
Phase II Accounting	4. Quantity 5. Quality 6. Benefit
Phase III Sequity	7. Equality 8. Conservation 9. Sefficiency
Phase IV Solution	10. Differential 11. Demand / Supply 12. Good enough alternative

Fig. 5.3 Phases, steps and feedbacks in water management

- Water quantities start with the justification of those WPTs (Fig. 2.1) that are zero or negligible (say less than 1% of Inflow). Then the rest of the WPTs, at least at the Meso level, is expanded into their WPIs, and a new schematic with all the WPIs is drawn, and their quantities are set (measured, calculated or estimated). They must obey the law of water balance as explained in the Theory and presented in the equations of the Sect. 4.1. It should be noted that the water balance must be obeyed at the level of Meso as well as Macro (if this level is part of the analysis). In this context, preliminary minimum and maximum values of water quantity needs for population, irrigation, industry, river, etc. should be specified.

- Water qualities are fundamental in advancing good water management for both people and nature. The development of quality weights for Sefficiency can start with justifying those W_{qX} values that are either zero or one, and subsequently setting the rest of the quality weights (measured, calculated or estimated). Then the minimum and maximum quality weights should be specified, which allow the possibility to manage and design systems with different water quality, particularly under water scarcity. Almost everywhere, water for direct human use must be treated obeying the specified min/max values of the quality parameters.

- Water benefits are, of course, the reasons that we use water for almost everything: drinking, health, irrigation, industry, spirituality, etc. To come up with beneficial weights, we also need to look at the costs attached to every WPI beginning with justifying the beneficial weights that are one e.g. water for human drinking (the highest priority). In estimating beneficial weights, also noise (Sect. 4.3) should be analysed in connection with WPIs.

The Sequity phase (III) is about the sustainable equity of system, which is comprised of advancing water equality and conservation as explained in the previous subsections, and Sefficiency as presented in Chapter 4.

The Solution phase (IV) also has three steps needed in a proper decision-making. They are differential, demand/supply, and good enough alternative. The differentials in Sequity lead us to equality, conservation and three Sefficiency impacts (see Sect. 4.4). They are crucial in leading water managers toward a sustainable, adaptive and efficient WUS. The outflow/inflow (demand/supply) values also change, particularly under water scarcity and climate change, and consequently they should be varied by going back to Phase II. This promotes some kind of sensitivity analysis with clear focus on more relevant WPTs, such as ET in irrigation systems or VA and its quality for urban areas. At the end, a solution should be reasonable and good-enough (Sect. 1.2) for the decision-makers and most of the stakeholders, particularly the poor and the less advantaged.

References

Burck J et al (2018) Climate change performance index: results 2019. German Watch, Bonn, Germany

Cambridge Dictionary (2019) Conservation. [Online] Available at: https://dictionary.cambridge.org/dictionary/english/conservation. Accessed 22 Aug 2019

Cambridge Dictionary (2019) Judgment. [Online] Available at: https://dictionary.cambridge.org/us/dictionary/english/judgment. Accessed 28 June 2019

Dellapenna JW (2001) The customary international law of transboundary fresh waters. Int J Glob Environ Issues 1(3–4):264–305

Dinar A, Pochat V, Albiac J (2015) Water pricing experiences and innovations. Springer International Publishing, Switzerland

European Parliament & Council (2000) Water Framework Directive, Official Journal L 327, European Union. [Online] Available at: http://ec.europa.eu/environment/water/water-framework/index_en.html. Accessed 20 March 2018

Fleurbaey M (2016) Economics and economic justice. [Online] Available at: https://plato.stanford.edu/archives/win2016/entries/economic-justice. Accessed 26 Aug 2019

Gleick PH (2010) Roadmap for sustainable water resources in southwestern North America. In: Proceedings of the National Academy of Sciences, 14 December, vol 107, issue 50

International Law Association (2004) Berlin rules on water resources. United Nations Economic Commission for Europe (UNECE), Geneva, Switzerland

Loucks D, Van Beek E (2005) water resources systems planning and management: an introduction to methods, models and applications. UNESCO, Paris

Oxford Dictionary (2019) Need. [Online] Available at: https://www.oxfordlearnersdictionaries.com/definition/english/need_1?q=need. Accessed 12 Oct 2019

Rawls J (1999) A theory of justice, Revised edn. Harvard University Press, Cambridge

Sen A (1992) Inequality reexamined. Oxford University Press, New York

Sen A (1999) Development as freedom. Alfred A Knope inc, sl

Sen A (2009) The idea of justice. Belknap Press of Harvard University Press, Cambridge, Massachusetts

Stone D (2012) Policy paradox: the art of political decision making, 3rd edn. W. W. Norton & Company Inc., New York

UNDP-HDR (2018) Human development indicators and indices 2018 statistical update: reader's guide. United Nations Development Programme, NY

UNESCO-WWAP (2006) WWDR2: world water development report 2. United Nations UNESCO and Berghahn Books, Paris and New York

UNGA (2010) Resolution 64/292: The human right to water and sanitation. [Online]. Available at: https://undocs.org/A/RES/64/292. Accessed 20 Oct 2019

UN-SDG (2019) The sustainable development goals report. United Nations, New York

UN-Water (2019) Water as a global issue. [Online]. Available at: https://www.un.org/en/sections/issues-depth/water. Accessed 10 Aug 2019

Chapter 6
Applications

This chapter is to help the readers appreciate the wide range of situations that the concepts and tools presented in this book can be employed in order to get a better insight into their functionings. Most of the examples given below are not real case studies with valid data; however, they are mostly adapted from real situations. Furthermore, the exercise is not to present full-fledged cases but rather to link a number of dots for a specific theme and give references for further studies. Finally, to show the input data and the results of Sefficiency, two templates are available: (a) a free MS-Excel tool at Haie (Sefficiency (Sustainable efficiency) of Water-Energy-Food Entangled Systems 2016) as a supplementary document that has the equations and a simple format to simulate various cases side by side, (b) Appendix B: Sefficiency Template is the compact form used for this book.

Note In all the tables of this chapter, grey cells display calculated numbers and the underlined numbers are the final solutions or values.

6.1 Water Saving Myth

There is a misconception between water supply/demand and water saving; often thinking that decreasing water supply will save water for downstream users. This may prove to be incorrect due to the law of water balance and the fact that water outflows have two states (FIW2; Sect. 2.4). This myth can be shown through a simple example using the concepts and symbols developed in this book.

© The Editor(s) (if applicable) and The Author(s), under exclusive license 91
to Springer Nature Singapore Pte Ltd. 2021
N. Haie, *Transparent Water Management Theory*,
Water Resources Development and Management,
https://doi.org/10.1007/978-981-15-6284-6_6

Table 6.1 The myth between water supply/demand and water saving, an example

Quantity	2010	Rise in demand %	Saving %	2030 base	2030 saving
I	200	50	15	300	255
C	40	50	0	60	60
R = RF	160			240	195
VU	400			400	400
VD	360			340	340

Let us assume that the amount of water that flows in a river is 400 units (VU). An urban area, in 2010, withdraws half of the water of the river (I) which 40 units is consumed (C) (Table 6.1). For simplicity, let us also assume that no leakage occurs in the water supply and waste water systems and RP = 0. Using Eq. 2.2, we can calculate total return flow to the river (R = RF = 200 – 40 = 160), and the flow of water downstream from the urban area (VD = 400 – 200 + 160 = 360).

Assuming a 50% rise in the urban demand of water by 2030, we have I = (1 + 0.5)*200 = 300, out of which 60 units are consumed. In this base scenario, R becomes 240 and VD = 340 units. However, planning a water conservation scenario for 2030 by promoting 15% water saving in water supply (I) without influencing water consumption, we get an abstraction of 255 units from the river with R = 195. This water saving scenario maintains 340 units of water after the urban area (VD), which is *equal* to the base scenario, i.e., without the water saving plan.

Hence, a water conservation plan did not save water for the downstream users of the urban area, but it resulted in reducing other factors, such as, energy consumption, equipment purchase, and water pollution. These are, of course, very important but from the water point of view, no water was saved, which is generally the declared objective. This is highly significant for the water scarce regions and it is an illusion thinking that water conservation plans will automatically result in more water in rivers and aquifers. Essentially real water saving is possible if the water managers reduce water consumption (C) of a WUS, but as we have seen throughout this book, the real question is how much WPIs (including consumptive ones) and their weights should change in order to achieve a more sustainable performance. Additionally, water development should focus on sectors of economy and ask which ones give higher Sefficiencies by also looking into TUF (Sect. 2.5) rather than only consumption.

6.2 Urban

Here, we apply Sefficiency to a simplified urban water cycle, which includes both water supply and wastewater systems. This means that the WUS under analysis with all its annual inflows and outflows (Fig. 2.1) is an urban water cycle, i.e., water in pipes, reservoirs, etc. Consequently, precipitation (PP) and evapotranspiration (ET) are practically zero. In addition, the urban area gets most of its water (VA) from a

Table 6.2 One to many Water Path Types

Symbol	Description
NRcit	City NR, which is the water consumed by the population of the city, such as, drinking and preparing food
NRind	Industrial NR, such as, virtual water and evaporation
RPirr	Water taken from the WUS for irrigation purposes (gardens and green zones)
RPwsl	Water supply system leaks, such as, leakage from the pipes that distribute clean water
RPwwl	Wastewater system leaks, such as, leakage from the pipes that collect wastewater

river (with a certain VU) and the rest from an aquifer (OS). Return to the river (RF) gives the downstream water (VD). However, NR and RP are different in that they have more than one WPI (i.e., they have a one to many relationship) as given in Table 6.2.

Figure 6.1 Shows the schematic of our urban water cycle example. The specific values of all the WPIs are summarised in Table 6.3 along with their quality and beneficial weights. Regarding this input data, let us briefly explain the reason for the W_b values of the two leakages that are generally considered to be non-beneficial, i.e., $W_b = 0$. All urban water cycles have leakage, which cannot be totally eliminated because there is a minimum quantity that is unavoidable real leakage (European Commission 2013; Leaks Suite 2019) and can be most of the total leakage of an urban system. If this unavoidable part has any value (e.g., goes to groundwater), then it should be integrated into the Sefficiency calculations with a positive W_b. With this in mind, Table 6.4 gives the Sefficiency values, i.e., the 3ME numbers. If water quantity for a WPI is zero, the values of the two weights are irrelevant. Finally, for this example there is no energy consideration (ENN $= 0$) but it will show up in the case for water-energy-food below.

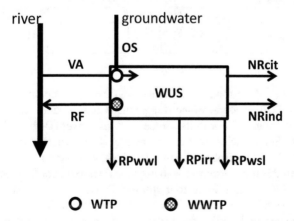

Fig. 6.1 Schematic of the urban water cycle WUS

Table 6.3 Input data for the urban example

WPI	Quantity	Weight Quality	Weight Beneficial	Usefulness Criterion
Evapotranspiration, ET	0	0	0	0
Non-reusable, NRcit	5,057,980	1	0.60	0.6
Non-reusable, NRind	5,057,980	1	0.60	0.6
Other Sources, OS	73,000	0.80	0.60	0.48
Total Precipitation, PP	0	0	0	0
Return Flow to source, RF	16,145,968	0.65	0.50	0.33
Other Return, RPirr	2,557,980	1	0.45	0.45
Other Return, RPwsl	8,526,600	1	0.10	0.1
Other Return, RPwwl	5,286,492	0.20	0.10	0.02
Abstracted water, VA	42,560,000	0.80	0.60	0.48
Downstream, VD	62,961,968	0.75	0.60	0.45
Upstream, VU	89,376,000	0.80	0.60	0.48
Water Balance, MesoSE	0.0			
Water Balance, MacroSE	0.0			
Energy, ENN	0			

Table 6.4 Sefficiency results for the urban example

ic = 1 (inflow)		ic = 0 (consumptive)		
Full Sefficiencies	%	Full Sefficiencies	%	
iMacroSE	85.0	cMacroSE	48.6	
iMesoSE	65.6	cMesoSE	46.3	
MicroSE	29.7	MicroSE	29.7	
Quantity Sefficiencies		Quantity Sefficiencies		
iMacroSEb	86.4	cMacroSEb	45.4	
iMesoSEb	65.2	cMesoSEb	40.5	
MicroSEb	23.7	MicroSEb	23.7	

The low values of consumptive Sefficiencies (ic $= 0$) are common in urban water cycles because TUF is generally much bigger than Consumption, even though around 80% of the Inflow goes back to its source, such as a river (Dworak 2007). Consequently, inflow Sefficiencies are more interesting and appropriate indicators, particularly, iMacroSE and iMesoSE. It seems that the first intervention for improving the performance of this urban example is to improve iMesoSE in order to go beyond say 80%. This problem has, at least, four specific dimensions: water supply, demand, leakage, and the influence of treatment plants in dealing with pollution. These four are interconnected with trade-offs, and the performance of an urban water cycle largely depends on the policies and decisions that set their operating regimes.

The Micro Sefficiencies are very low because they are of the Classical Efficiency type. In this example, Eq. 4.16 gives a very low CE value of 36% illustrating what we wrote in showing that it is flawed (Sect. 4.5). However, some experts use only water quantities to arrive at a meso level efficiency, which in this case is about 68% ($= (I - RP_{wsl} - RP_{wwl})/I$). Such calculation ignores the Usefulness Criterion and is a fraction with very limited meaning, at least in relation to the full performance of the system. Furthermore, it implicitly assumes that the beneficial and quality weights of all the involved flows are practically the same (e.g., $W_{sX} = 1$), which is, of course, a wrong assumption and, consequently such a calculation that is based only on water quantity should not be used.

6.3 Equity

Equity in urban areas is one of the critical issues that the world faces. As stated in Sect. 5.5, any system during a time interval is defined by a point on one of the 15 segments shown in Fig. 5.1. One important fact about this figure worth repeating is that local managers in consultation with stakeholders can largely set the thresholds (Sect. 5.1). There is also a necessity to have some water in river after abstraction (i.e., after VA) and after return (VD) due to ecological needs, legal contracts, and ethics (national or international). Such a downstream required value is shown as DS_{req} and its minimum as $DS_{req, min}$. Now, let us assume that G in Fig. 6.2 is an urban area with Table 6.5 showing its input data (L^3/T = volume of water per unit of time). They exhibit the following two issues:

- Water inequality (Sect. 5.3) because M2 is below Amin and M1 is clearly above Min
- Downstream from the point of water abstraction (VA), the river does not have enough water to satisfy the stated minimum requirement ($DS_{req, min}$)

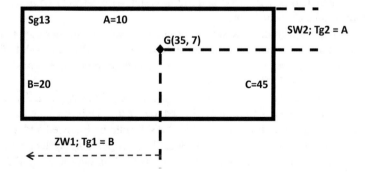

Fig. 6.2 Urban area with water shortages and excesses

Table 6.5 Input data for the urban example of equality and conservation

Item	L3/T/person	L3/T	Comments
M1 population		800,000	
M2 population		200,000	
Total population		1,000,000	
VU		30,000,000	
DSreq, min		3,000,000	minimum downstream required
Amin threshold	10	2,000,000	Target A for M2
Min threshold	20	16,000,000	Target B for M1
Min threshold	20	4,000,000	Target B for M2
Max threshold	45	36,000,000	Target C for M1
Amax threshold	48	38,400,000	for M1
M1	35	28,000,000	
M2	7	1,400,000	Sg13; PtII
M1+M2 (VA)		29,400,000	
downstream after VA		600,000	downstream after VA is not ok
VD		24,120,000	assuming 80% return (RF)

To solve these problems, we recognize that G is in segment 13, which belongs to Policy type II – PtII (Table 5.6). Table 6.6 presents three steps to highlight the processes that lead to solving the two issues. The numbers are demonstrative and the comments on them are self-explanatory having in mind the subsection in Chap. 5 on PtII. The following points are also helpful in understanding the table.

- There are two different Min thresholds in the volume column per time ($L^3 T^{-1}$) because M1 and M2 population are different.
- Reaf is first calculated using Eq. 5.1) and then should be adjusted in a reflection meeting with the stakeholders (Sect. 3.2).
- First step takes M2 out of the Extreme Water Poverty with a little help from M1, which can happen by reallocating 5% of its excess water.
- Second step takes M2 out of Water Poverty into Water Wise by reallocating 17.6% of M1 water. This puts both M1 and M2 in Water Wise category with M1 still in a more comfortable condition.
- Third step deals with the downstream problem. In solving this question the Reaf that is assumed should be greater than or equal to $(VA + DS_{req, min} - VU)/ZW1$. In reality and as was explained (Sect. 5.6) its value depends on the consensus of water managers and M1 population.
- In this simple and widespread problem, the final solution satisfies $DS_{req, min}$ and allocates about 23 and 4 million to M1 and M2, respectively, pushing the system toward equality. Since both are in Sg33, PtIV should be followed and the Seffi-ciency of the urban WUS analysed (Sect. 1.2). In pursuing system efficiency, M2 and downstream water should not go below the Min threshold and $DS_{req, min}$, respectively.

Table 6.6 Results for the urban example of equality and conservation

Item	Step 1 Equality	Comments	Step 2 Equality	Comments	Step 3 Conservation (and Equality)	Comments
Tg1 min	16,000,000	Target B; Table 9	16,000,000	Target B; Table 9	16,000,000	Target B; Table 9
Tg2	2,000,000	Target A; Table 9; Priority	4,000,000	Target B; Table 9; Priority	4,000,000	Target B; Table 9
SW2	600,000		2,000,000		0	
ZW1 max	12,000,000		11,400,000		9,393,600	
Reaf, min	0.050	with NEW=0	0.175	with NEW=0	0.000	with NEW=0
Reaf assumed	0.050		0.176		0.256	depends on M1
Reaf * ZW1	600,000	reallocation to M2	2,006,400	reallocation to M2	2,404,762	conservation
NEW	0	NEW not needed	0	NEW not needed	0	NEW not needed
NEW+Reaf*ZW1-SW2	0	M2 gets to target	6,400	M2 gets to target	2,404,762	
M1	27,400,000		25,393,600		22,988,838	
M2	2,000,000	Sg23; PtII	4,006,400	Sg33; PtIV	4,006,400	Sg33; PtIV
M1+M2 (VA)	29,400,000		29,400,000		26,995,238	
downstream after VA	600,000	downstream after VA is not ok	600,000	downstream after VA is not ok	3,004,762	downstream after VA is ok

6.4   Farm

In most parts of the world, crop growth consumes most of the available freshwater in a region, hence making its efficient use a top priority. Assuming a one to one relationship between WPTs and WPIs, Fig. 2.1 (without NR) becomes the schematic for this example, with Table 6.7 giving the input data and Table 6.8 the 3ME results.

The Sefficiencies seem acceptable but can be improved by first focusing on cMacroSE, which is the ratio of useful Consumption (C_s) to Macro-TUF_s (Sect. 4.2). To do this, measures should be adopted that relatively increase the numerator of cMacroSE—Eq. 4.11—more than the denominator. This can be done in many ways by changing the three Pillars identified in Table 6.7. For example, in the region, the developmental priorities of various economic activities can be reformed, hence changing

Table 6.7 Input data for the agricultural example

WPI	Quantity	Weight Quality	Weight Beneficial	Usefulness Criterion
Evapotranspiration, ET	240,000	1	0.7	0.7
Non-reusable, NR	0	0	0	0
Other Sources, OS	5,000	0.7	0.7	0.49
Total Precipitation, PP	5,000	1	0.15	0.15
Return Flow to source, RF	50,000	0.4	0.6	0.24
Other Return, RP	80,000	0.4	0.6	0.24
Abstracted water, VA	360,000	0.8	0.8	0.64
Downstream, VD	190,000	0.55	0.65	0.36
Upstream, VU	500,000	0.8	0.8	0.64
Water Balance, MesoSE	0.0			
Water Balance, MacroSE	0.0			
Energy, ENN	0			

Table 6.8 Sefficiency results for the agricultural example

ic = 1 (inflow)		ic = 0 (consumptive)	
Full Sefficiencies	%	Full Sefficiencies	%
iMacroSE	78.9	cMacroSE	71.2
iMesoSE	85.3	cMesoSE	83.0
MicroSE	71.9	MicroSE	71.9
Quantity Sefficiencies		Quantity Sefficiencies	
iMacroSEb	84.0	cMacroSEb	72.2
iMesoSEb	84.2	cMesoSEb	78.4
MicroSEb	57.5	MicroSEb	57.5

W_{bX} values in such a way that favours higher values for the four full Sefficiencies. As also mentioned in FIW4a, water managers can act as social catalysts for change in the direction of sustainable development. A concrete scenario assumes that this agricultural area is in a water scarce region and we may want to decrease Consumption (ET) by 10% (W_{bET} stays the same) and decrease water allocated to the farm (VA) by the same quantity. This is a rather prevalent idea among the water experts and professionals, but in our case, this leads to lower performance for the WUS, for example, making cMacroSE worse off by about 5p.p (i.e., 66.5%). This, of course, should not be accepted. However, there is the idea that the decrease in ET is mostly due to high decrease in evaporation, and moderate increase in transpiration (hence yield), which causes an increase in the benefits for the farmer, i.e., an increase in W_{bET}. This seemingly simple and attractive statement is actually very complex. For example, if we increase W_{bET} from 0.7 to 0.72 (about 3% increase in benefits), we still get a lower cMacroSE of about 3p.p, which again makes the decrease in ET unacceptable. Furthermore, by decreasing VA, we get less RP and RF, which in turn make trade-offs less obvious, particularly if RP was recharging groundwater. Additionally, if the system is for drip irrigation, there are the problems presented by Burt et al. (Irrigation Performance Measures—Efficiency and Uniformity 1997) and others (also see FIW1a). Definitely we are not against drip systems but simply highlighting the very complex nature of the trade-offs as can be verified by their patterns presented in Sect. 4.4. Of course, as mentioned in the first example of this Chapter, real water saving occurs only by decreasing Consumption, but this is acceptable if Sefficiency stays the same, increases or remains Green (Sect. 5.3).

Equation (4.16) gives a CE value of 46.5%, which is closer to MicroSE$_b$ of 57.5%. However, its very low value relative to Meso and Macro is again another proof of the flaw of CE (Sect. 4.5). Of course, those who utilize CE have a louder voice in our institutions and organisations for urgent investments, but years after implementation it may prove to be detrimental to the development of the (water scarce) region.

6.5 Water, Energy, Food

Water-energy-food (WEF) nexus is intensifying due to water scarcity and pollution, energy pollution and food insecurity. There are various reasons for these issues, such as, population, climate, economy, technology and governance. Here, we present an example by adopting some of the material from Haie (Sefficiency (Sustainable efficiency) of Water-Energy-Food Entangled Systems 2016) including the schematic of our WUS—Fig. 6.3, which is a simple one to one relationship of Fig. 2.1.

WEF is a complex system of systems with many changing parts. In order to analyse it in terms of water and using Sefficiency to quantify its performance, we need to associate food and energy to the three Pillars of a WUS as shown in Fig. 6.3 and explained below.

We know that as crops have better access to more water (within limits), they transpire more and consequently produce more yield (Perry et al. 2009). Most of ET

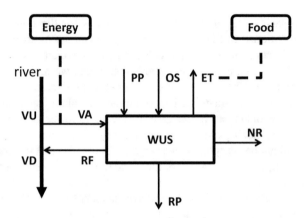

Fig. 6.3 Typical schematic for water–energy–food (WEF) nexus (Haie (Sefficiency (Sustainable efficiency) of Water-Energy-Food Entangled Systems 2016) by permission of Taylor & Francis Ltd, http://www.tandfonline.com)

is transpiration, which makes (crop) ET a direct and proper proxy for representing food. It should be noted that like everything else ET has limits (FIW4c) and depends on water as well as other factors, but from the point of view of Sefficiency, ET is one of the WPTs of the WUS under analysis.

However, energy and more precisely energy costs influence the beneficial Pillar of a WUS. This cost has more than one dimension, however from the point of view of a farmer/producer, such costs reduce the profits. For example, as a producer pumps more water to produce more crop, the net profit changes, which has non-linear effects on the performance of the system. To quantify it, we should convert energy cost to some form of energy factor (varying between 0 and 1), which negatively influences the beneficial weights of the relevant WPIs. To find those factors various methods can be employed, the simplest being $ENN = (EN - EN_{min})/(EN_{max} - EN_{min})$, with EN_X being a (stepwise) energy cost as a function of a WPI, and ENN_X a normalized energy function between 0 and 1 as given in Eq. (6.1).

$$EN_X = f(X)$$
$$ENN_X = f(EN_X), 0 \leq ENN_X \leq 1 \qquad (6.1)$$

For this example, considering $W_{bET}1$ and $W_{bET}2$ as the weights without and with energy, respectively, we get $W_{bET}2 = (1 - ENN_{VA}) * W_{bET}1$. Furthermore, we chose linear functions for both EN and ENN as follows: $EN = 4.41$ VA $- 352.9$ and $ENN = 0.0013$ EN with more details at Haie (2016) or its preprint at my Google site (Haie, 2020). Table 6.9 gives the input data and Table 6.10 the 3ME values for this example.

Table 6.10 reveals that among the four significant Sefficiencies iMacroSE came out to be acceptable. However, cMacroSE and cMesoSE as well as iMesoSE should be the focus of improvement, which need various changes to the WEF systems. One thing worth noticing is that the energy cost is steep, which makes much impact on

Table 6.9 Input data for the water-energy-food example

WPI	Quantity	Weight		Usefulness
		Quality	Beneficial	Criterion
Evapotranspiration, ET	65	1	0.69	0.69
Non-reusable, NR	0	0	0	0
Other Sources, OS	0	0	0	0
Total Precipitation, PP	0	0	0	0
Return Flow to source, RF	35	0.50	0.45	0.225
Other Return, RP	20	0.50	0.45	0.225
Abstracted water, VA	120	0.80	0.90	0.72
Downstream, VD	215	0.72	0.90	0.65
Upstream, VU	300	0.80	0.90	0.72
Water Balance, MesoSE	0.0			
Water Balance, MacroSE	0.0			
Energy, ENN	0.23			

Table 6.10 Sefficiency results for the water-energy-food example

ic = 1 (inflow)		ic = 0 (consumptive)	
Full Sefficiencies	**%**	**Full Sefficiencies**	**%**
iMacroSE	87.3	cMacroSE	62.1
iMesoSE	66.2	cMesoSE	60.6
MicroSE	51.9	MicroSE	51.9
Quantity Sefficiencies		**Quantity Sefficiencies**	
iMacroSEb	91.6	cMacroSEb	66.4
iMesoSEb	64.4	cMesoSEb	53.9
MicroSEb	41.5	MicroSEb	41.5

the performance of the system. We should also analyse the results by looking into the three impact categories mentioned in Sect. 4.4:

- The I/O impacts are high (more than 5p.p.)
- The level impacts are high except for the difference between cMacroSE and cMesoSE that is negligible (less than 2p.p.)
- The pollution impacts at Macro are medium (between 2 and 5p.p.), at iMeso is negligible, and at cMeso is high

This is a situation that there are various high impacts, necessitating a fundamental change in water (re)allocation plans and pollution control. For more ideas, the reader can use the Sefficiency template (freely available from the above-mentioned paper) to simulate the system with different energy costs and read about a parallel example with four scenarios and discussions.

6.6   River, Urban, Farm

We have been clear about the difference between water resource and water use (Sect. 2.1). In this case, we present a river section as a WUS that receives water from upstream (inflows) and supplies water (outflows) to an urban area and an irrigated region while guaranteeing a predefined minimum amount of water for downstream. This example is taken from the Open Access paper of Ahmad and Haie (2018), which analyses the situation of the Kano River in Nigeria (an upper tributary of lake Chad) and the impacts of population growth and climate change until 2050. Here, we describe some general aspects and invite those interested to refer to the paper for more details.

According to Fig. 5.3, after defining the WUS and its objectives, we should draw the schematic trying to follow the flows in achieving water balance. This is done for this case and shown in Fig. 6.4 with all the Water Path Instances (WPIs), which we do not describe as they are self-explanatory (also see Table 2.1) and can be verified in the said paper.

Writing the water balance equation for Fig. 6.4 is easy, however to get somewhat accurate data is generally difficult, and for the Kano River proved to be very difficult. These are situations that the approach described in Chap. 3 is worth pursuing. For the

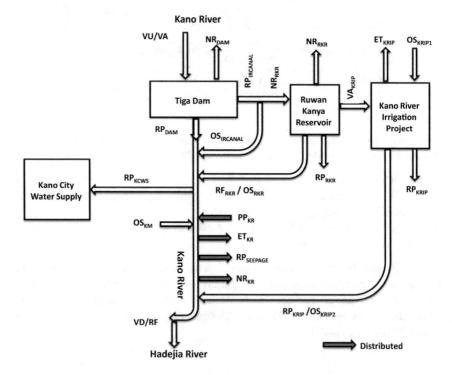

Fig. 6.4   Schematic of the WUS of Kano River showing all the WPIs (Ahmad and Haie 2018)

case presented here, questionnaires were developed and two groups, namely, farmers and managers were consulted and the results helped to set the numbers assumed for various variables of the Sefficiency indicators. These crude numbers can serve as initial data that through years can get closer to reality, particularly if smart data gathering technologies are used.

Finally, allocating water to multiple users under uncertainty should be the norm. To develop scenarios, various models, including probabilistic, stochastic and fuzzy methods have been used. For example, Loucks and Van Beek (2005) utilize the Monte Carlo simulation in order to distribute the water of a river among three users according to a number of policies, including a minimum amount of water in the river. In such cases, Sefficiency can be applied as explained for Kano River in order to understand the performance of all the allocations together.

6.7 Trade and Water Footprint

The idea of Water Footprint (WF) was first coined by the late Hoekstra (2003) and then the Water Footprint Network (WFN) developed a methodology in order to implement it in various circumstances (Hoekstra et al. 2011). WFN defined footprint as a volumetric metric that was different from its traditional meaning of impact metric that had been employed by the life-cycle assessment (LCA) community, which later developed its own water footprint guidelines (ISO 2014) in order to help LCA in its impact analyses.

Throughout the last decade, many developments made WF globally famous hence attracting criticisms, which in the most part were not complete nor consistent. Haie et al. (2018) refuted almost all those criticisms in six categories and employed Sefficiency to highlight a solution for the third phase of the WF methodology - sustainability assessment (Hoekstra et al. 2011). To demonstrate our solution, we utilized an example put forth by one of the water economists in criticising WF. This is briefly presented here and we encourage those interested to go to Haie et al. (2018), which is freely available for download.

According to WFN, WF is the sum of blue, green and/or grey water footprints (Hoekstra et al. 2011). In Sefficiency terminology, blue and green WF is Consumption and grey WF may be assumed to be R_{nq} (Sect. 2.5). If this assumption is reasonable it means that $WF = C + (1 - W_{qR}) R = I - R_q$. Comparing this expression with that of TUF_q of Eq. 2.5) we notice that in practice WF is greater than TUF_q, which we propose to be the real value of water footprint. The reason for this discrepancy is that WF, contrary to TUF_q, was not developed through a foundational theory and a consequent solid approach as was done in this book.

For this example, data for grey water is not given making WF equal to Consumption while ignoring pollution. The example compares countries A and B (Fig. 6.5) that trade cotton and wheat. Country A is relatively water scarce and country B has a worse climate. The irrigation requirements of the crops and their output prices are

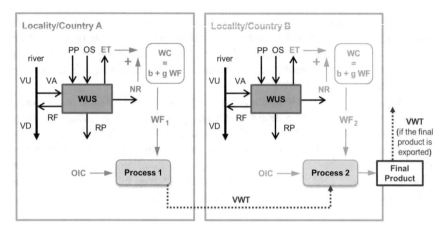

Fig. 6.5 Schematic of WUS, water footprint and trade between two countries (Haie et al. 2018); (b + g) WF: blue + green water footprint; WC: water consumption; VWT: virtual water trade; OIC: other input consumption; all other variables in Table 2.1)

different. Please refer to the 2018 reference mentioned above for assumptions, data, results and discussions for three scenarios.

References

Ahmad M, Haie N (2018) Assessing the impacts of population growth and climate change on performance of water use systems and water allocation in Kano River basin, Nigeria. Water 10(12):1–21

Burt C et al (1997) Irrigation performance measures—efficiency and uniformity. J Irrigation Drain Eng ASCE 123(6):423–442

Dworak T et al (2007) EU water saving potential (Part 1 - Final Report). Ecologic Institute for International and European Environmental, Berlin

European Commission (2013) Resource and economic efficiency of water distribution networks in the EU, Final Report. European Commission, Brussels, Belgium

Haie N (2016) Sefficiency (Sustainable efficiency) of water-energy-food entangled systems. Int J Water Resour Dev 32(5):721–737

Haie N (2020) My Google website. [Online] Available at: https://sites.google.com/view/naimhaie5. Accessed 13 April 2020

Haie N, Freitas MR, Pereira JC (2018) Integrating water footprint and sefficiency: overcoming water footprint criticisms and improving decision making. Water Alternatives 11(3):933–956

Hoekstra AY (2003) Virtual water trade. Delft, The Netherlands, sn

Hoekstra A, Chapagain A, Aldaya M, Mekonnen M (2011) The water footprint assessment manual: setting the global standard. Earthscan, London

ISO (2014) ISO 14046: environmental management: water footprint—principles, requirements and guidelines, 1st edn. International Organization for Standardization, Geneva, Switzerland

Leaks Suite (2019) Unavoidable annual real losses & infrastructure leakage index [Online]. Available at: https://www.leakssuitelibrary.com/uarl-and-ili. Accessed 27 Dec 2019

Loucks D, Van Beek E (2005) Water resources systems planning and management: an introduction to methods, models and applications. UNESCO, Paris

Perry C, Steduto P, Allen RG, Burt CM (2009) Increasing productivity in irrigated agriculture: agronomic constraints and hydrological realities. Agric Water Manage 96:1517–1524

Appendix A
Equivalency

Proof of the equivalency of minimizing a difference and maximizing its ratio (Haie and Keller 2012):

Let Z1 be the minimum of the difference (F − H) as follows:

- Z1 = min (F − H), with F and H positive

The solution (i.e., real values of F and H) is the minimum value of F and the maximum value of H. These two values of F and H yield the minimum for (F − H), or Z1 = min (F) − max (H).

Now let Z2 be the maximum of the ratio (H/F) as follows:

- Z2 = max (H/F)

The solution is the minimum value of F and the maximum value of H. Therefore, although Z1 and Z2 are different, the two solutions (i.e., the values for F and H) are the same.

Reference

Haie N, Keller A (2012) Macro, meso, and micro-efficiencies in water resources management: a new framework using water balance. J Am Water Resour Assoc (JAWRA) 48(2):235–243

N. Haie, *Transparent Water Management Theory*,
Water Resources Development and Management,
https://doi.org/10.1007/978-981-15-6284-6

Appendix B
Sefficiency Template

A Sefficiency template in MS-Excel with cell formulas is available freely as a supplement of Haie (2016) paper, which shows a one to one mapping of WPTs and WPIs. Table B.1 gives the Sefficiency template for input data utilized in this book, and Table B.2 shows the 3ME Sefficiency values. Both of these tables are compact forms of the Excel template. In these tables, the grey cells display numbers that are calculated and the underlined numbers (now empty) are the final solutions or values.

Reference

Haie N (2016) Sefficiency (sustainable efficiency) of water-energy-food entangled systems. Int J Water Resour Dev 32(5):721–737

N. Haie, *Transparent Water Management Theory*,
Water Resources Development and Management,
https://doi.org/10.1007/978-981-15-6284-6

Table 1 Compact Sefficiency Template (input data)

WPI	Case 1	Weight		Usefulness
	Quantity	Quality	Beneficial	Criterion
Evapotranspiration, ET				
Non-reusable, NR				
Other Sources, OS				
Total Precipitation, PP				
Return Flow to source, RF				
Other Return, RP				
Abstracted water, VA				
Downstream, VD				
Upstream, VU				
Water Balance, MesoSE				
Water Balance, MacroSE				
Energy, ENN				

Table 2 Compact Sefficiency Template (3ME)

ic = 1 (inflow)		ic = 0 (consumptive)	
Full Sefficiencies	**%**	**Full Sefficiencies**	**%**
iMacroSE		cMacroSE	
iMesoSE		cMesoSE	
MicroSE		MicroSE	
Quantity Sefficiencies		**Quantity Sefficiencies**	
iMacroSEb		cMacroSEb	
iMesoSEb		cMesoSEb	
MicroSEb		MicroSEb	

Appendix C
Environment and Social Contract

This appendix is different from the rest of the book, as it tries to highlight probably the most important problem that causes environmental degradation. First, let us remember that water is a significant part of environment as is also obvious in its definition "the air, water, and land in or on which people, animals, and plants live" (Cambridge Dictionary 2019). On the other hand, there is no need to remind the readers that we are living under unprecedented social and environmental problems. Drivers, such as population, climate, economy, governance and technology are making sustainable water development as one of the major challenges of this century. Since 2015, the yearly Global Risks Report of the World Economic Forum presents "water crises" as the number one societal risk in terms of impact. Its 2019 edition (World Economic Forum 2019) puts "water crises" as the number two societal risk in terms of likelihood. Humanity has finally come to realize that fresh water is limited and must be dealt with as a scarce resource economically, socially and environmentally.

One way of looking into such environmental and social problems is to look into one of the central organising principles of human societies, the social contract theory. It "incorporates Enlightenment conceptions of the relation of the individual to the state. According to the general social contract model, political authority is grounded in an agreement (often understood as ideal, rather than real) among individuals, each of whom aims in this agreement to advance his rational self-interest by establishing a common political authority over all" (Bristow 2017).

The eighteenth century Enlightenment put forth the social contract theory, which is indeed a positive political and moral principle but it has serious limitations for our rapidly changing societies. The consequences of those shortfalls are readily visible in poverty, disadvantaged people (including women), health, lack of consideration for the future generation, etc. However, the most serious inadequacy of the policies according to the social contract model is their incompleteness: "Another structural problem of the social contract is related to the nationalistic presuppositions of the doctrine. In other words, the social contract is confined to the borders of a nation-state,

N. Haie, *Transparent Water Management Theory*, Water Resources Development and Management, https://doi.org/10.1007/978-981-15-6284-6

and does not include *all living human beings*... [I]t remains a fact that the nationalistic construction and definition of social contract is one of the most important causes of environmental waste and degradation. As long as the interests of a limited part of the world can be secured at the expense of the interests of other parts of the world, environmental destruction in the form of internal consumption and export of the environmental costs to other parts of the world will remain the guiding principle of policy making, resulting in both environmental destruction and social injustice" (Saiedi 2017).

References

Bristow W (2017) Enlightenment. [Online] Available at: https://plato.stanford.edu/archives/fall2017/entries/enlightenment. Accessed 13 Dec 2019

Cambridge Dictionary (2019) Environment. [Online] Available at: https://dictionary.cambridge.org/dictionary/english/environment. Accessed 13 Dec 2019

Saiedi N (2017) A Bahá'í approach to the environment. [Online] Available at: www.researchbahai.com. Accessed 3 Aug 2017

World Economic Forum (2019) The global risks report 2019, 14th edn, World Economic Forum, Geneva

Index

© The Editor(s) (if applicable) and The Author(s), under exclusive license
to Springer Nature Singapore Pte Ltd. 2021
N. Haie, *Transparent Water Management Theory*,
Water Resources Development and Management,
https://doi.org/10.1007/978-981-15-6284-6

Printed in the United States
by Baker & Taylor Publisher Services